AF410864

Energy Metabolism and Diet

In the case of complex lipids (Figure 7A), the analysis revealed a distinct separation of both krill oil-supplemented mice and mice given omega-3 PUFAs as TAGs from LHF-fed controls within the first Component (C1), which describes 63% of the total variation between these groups. Moreover, both groups fed with krill oil (i.e., ω3PL, ω3PL-R) separated from the ω3TG-R group (omega-3 PUFAs as TAGs) within C2 (Figure 7A). A separate analysis of polar metabolites (Figure 7B) showed a weaker separation of mice fed with krill oil from both LHF controls and mice fed a diet containing omega-3 PUFAs as TAGs within C1, which accounted for 28% of the total variation between the LHF and ω3TG-R groups on the one hand and the ω3PL and ω3PL-R groups on the other. Thus, krill oil supplementation had major effects on both complex lipids and polar metabolites, with omega-3 TAGs affecting mainly the lipidome. Subsequent VIP analysis indicated that several TAG species containing DHA and/or EPA (e.g., TAG 60:13; TAG 18:2_20:5_22:6, TAG 56:9; TAG 18:2_18:2_20:5 or TAG 60:12; TAG 18:2_20:4_22:6) were the most discriminating factors in terms of complex lipids (Figure 7C), while alkaloids stachydrine and trigonelline, as well as trimethylamine N-oxide (TMAO), represented the most discriminating factors among polar metabolites (Figure 7D). In the case of alkaloids, their concentration was significantly increased in the liver of ω3PL and ω3PL-R mice compared to the LHF and ω3TG mice (see left panels in Figure 8A,B), which is due to the increased concentration of these substances in the krill oil-containing diet (right panels in Figure 8A,B).

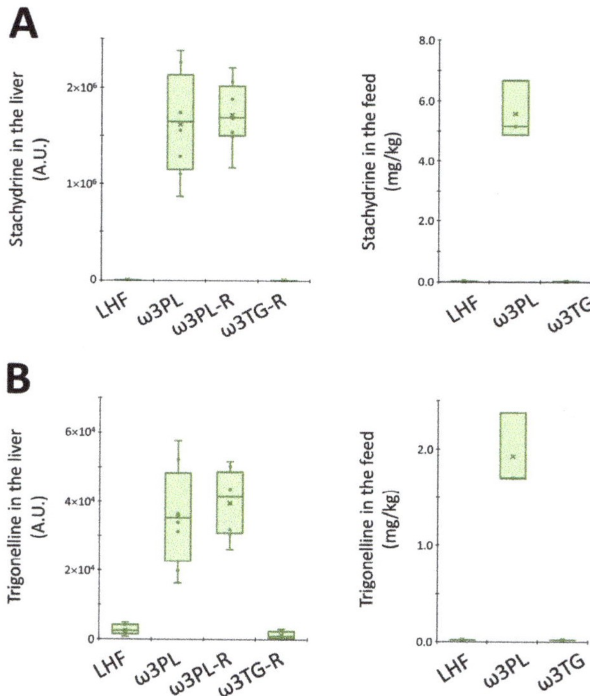

Figure 8. Box plots for stachydrine (**A**) and trigonelline (**B**) levels in the liver (arbitrary units; left panel) and in experimental diets (mg/kg; right panel).

Furthermore, we also performed a subanalysis of certain lipid classes with a known relationship to insulin resistance (e.g., diacylglycerols; DAGs) or DNL (e.g., short/medium-chain TAGs containing 38 to 48 carbons and 0 to 3 double bonds; see Section 4 for details) in the liver (Figure 9). We found that krill oil administration reduced the total DAG levels in liver tissue of ω3PL and ω3PL-R mice (Figure 9A), and in particular the content of perinsulinemic conditions (Figure 6F). Therefore, the above data suggest that the potent antisteatotic effects of krill oil supplementation in the livers of mice with diet-induced obesity and exacerbated hepatic steatosis are associated with improved whole-body and tissue sensitivity to insulin, but cannot be explained by changes in VLDL-TAG secretion.

3.5. Hepatic Metabolome in Relation to Tissue TAG Accumulation and Insulin Sensitivity

To better understand the underlying mechanisms of the potent antisteatotic and insulin-sensitizing effects of krill oil supplementation in mice with exacerbated hepatic steatosis, and to determine how these mechanisms differ in mice given omega-3 PUFAs as TAGs (see also Figure 1A), we analyzed the metabolipidomic profiles of the liver using four different LC-MS platforms (see Section 2.11 for details). First, the annotated data were analyzed using PLS-DA, supervised classification technique, to gain a general view on the impact of omega-3 PUFAs supplementation on both complex lipids and polar metabolites (Figure 7; see Table S7 for a complete list of annotated metabolites).

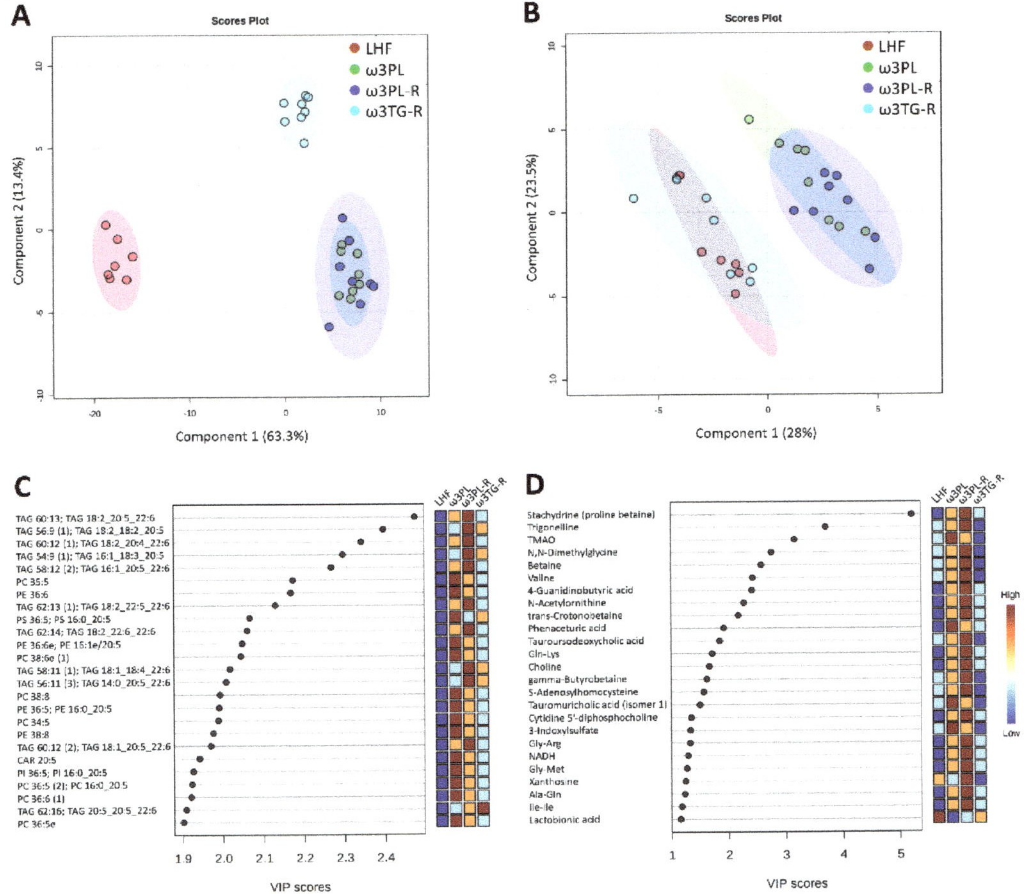

Figure 7. Four-class PLS-DA score plots of complex lipids (**A**; $n = 507$) and polar metabolites (**B**; $n = 157$) in the liver in response to dietary challenges, and the most discriminating complex lipids (**C**) and polar metabolites (**D**) based on VIP scores from PLS-DA.

Energy Metabolism and Diet

Editors

Arie Nieuwenhuizen
Evert van Schothorst

MDPI • Basel • Beijing • Wuhan • Barcelona • Belgrade • Manchester • Tokyo • Cluj • Tianjin

Editors
Arie Nieuwenhuizen
Human and Animal Physiology
Wageningen University
Wageningen
The Netherlands

Evert van Schothorst
Human and Animal Physiology
Wageningen University
Wageningen
The Netherlands

Editorial Office
MDPI
St. Alban-Anlage 66
4052 Basel, Switzerland

This is a reprint of articles from the Special Issue published online in the open access journal *Nutrients* (ISSN 2072-6643) (available at: www.mdpi.com/journal/nutrients/special_issues/diet_metabolism).

For citation purposes, cite each article independently as indicated on the article page online and as indicated below:

LastName, A.A.; LastName, B.B.; LastName, C.C. Article Title. *Journal Name* **Year**, *Volume Number*, Page Range.

ISBN 978-3-0365-1852-7 (Hbk)
ISBN 978-3-0365-1851-0 (PDF)

Contents

About the Editors

Arie Nieuwenhuizen

Arie G. Nieuwenhuizen is an assistant professor in human physiology at Human and Animal Physiology (HAP) of Wageningen University, Wageningen, The Netherlands. His main research topic is energy metabolism in human health and disease, with specific reference to skeletal muscle tissue and including influences of nutrition, exercise and aging. In this respect, the non-invasive techniques to assess energy metabolism in humans have his special interest, resulting in the successful application of techniques such as indirect calorimetry, analyses of volatile organic compounds in exhaled air and near-infrared spectroscopy to assess skeletal muscle mitochondrial capacity. He has published more than 70 scientific papers and co-organizes a post-graduate course "Energy Metabolism & Body Composition".

Evert van Schothorst

Evert M. van Schothorst is an associate professor in molecular physiology at Human and Animal Physiology (HAP) of Wageningen University, Wageningen, The Netherlands. His main research topics are energy metabolism from the whole body to molecular mechanisms, especially in the field of adipose tissue and insulin resistance, using nutrients and food bioactive compounds to improve white adipose tissue function in preclinical models. More recently, he focussed specifically on nutritional programming from the post-weaning period onwards by carbohydrates. This also provided him the means to extend the standard indirect calorimetry system to not only measure non-invasively and real-time energy metabolism (energy expenditure and substrate usage), but also to be the first to measure gut microbiota fermentation activity by measuring methane and hydrogen gasses. Furthermore, he incorporated 13CO2 and 12CO2 sensors, which enables measuring fluxes using 13C-labeled substrates. He has published over 90 scientific papers. He is a member of the European Association for the Study of Diabetes (EASD), secretary and member of the Dutch Association for the Study of Obesity (NASO), and co-organizer of post-graduate courses focussing on 'Metabolic programming', 'Functional and omics analysis of carotenoid interventions', and 'Temperature and Metabolism: Implications for nutritional studies'.

Preface to "Energy Metabolism and Diet"

We have great pleasure in presenting this Special Issue of *Nutrients* on "Energy metabolism and diet". The scope of this issue is the complex interaction between diet, energy metabolism and health or physical functioning, and all of the contributions to this issue aim to increase our understanding of this interaction and how it can be applied to improve human health and, more specifically, metabolic health. This Special Issue is particularly valuable to any scientist or health professional working in the field of health, nutrition and exercise. It is acknowledged that this valuable contribution to this field of expertise could only be realized thanks to the high-quality contributions of the authors Ciara Cooney, Ed Daly, Maria McDonagh, Lisa Ryan, Emmanuel Rineau, Naïg Gueguen, Vincent Procaccio, Franck Geneviève, Pascal Reynier, Daniel Henrion, Sigismond Lasocki, Gabriella Sistilli, Veronika Kalendova, Tomas Cajka, Illaria Irodenko, Kristina Bardova, Marina Oseeva, Petr Zacek, Petra Kroupova, Olga Horakova, Karoline Lackner, Amalia Gastaldelli, Ondrej Kuda, Jan Kopecky, Martin Rossmeisl, Rieneke Terink, Renger F. Witkamp, Maria T. E. Hopman, Els Siebelink, Huub F. J. Savelkoul, Marco Mensink, Jiri Funda, Radek Pohl, Tomas Cajka, Michal Hensler, Petra Janovska, Katerina Adamcova, Lucie Lenkova, Petr Zouhar, Pavel Flachs and Jerry Colca. Moreover, we are grateful to the editorial office of *Nutrients* for the assistance during the initiation and completion of this special issue.

Arie Nieuwenhuizen, Evert van Schothorst
Editors

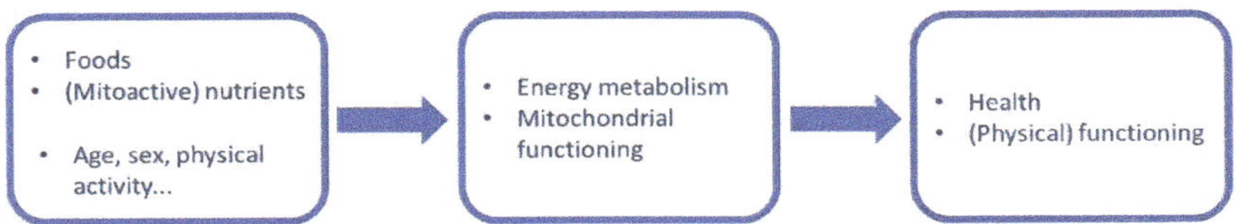

Editorial

Energy Metabolism and Diet

Arie G. Nieuwenhuizen * and **Evert M. van Schothorst**

Human and Animal Physiology, Wageningen University, 6708 WD Wageningen, The Netherlands; evert.vanschothorst@wur.nl
* Correspondence: arie.nieuwenhuizen@wur.nl; Tel.: +31-317482760

Energy metabolism at whole body and cellular, and even organelle (i.e., mitochondrial), level requires adequate regulation in order to maintain or improve (metabolic) health. In eukaryotic cells, mitochondria are key players in energy (ATP) production via oxidative phosphorylation. Both macro- and micronutrients potentially influence energy metabolism and mitochondrial functioning, either as substrates for (oxidative) catabolism or as essential constituents of enzymes or protein complexes involved in (mitochondrial) energy metabolism (Figure 1).

Figure 1. Schematic concept on the interaction between diet, energy metabolism, and health.

Citation: Nieuwenhuizen, A.G.; van Schothorst, E.M. Energy Metabolism and Diet. *Nutrients* **2021**, *13*, 1907. https://doi.org/10.3390/nu13061907

Received: 25 May 2021
Accepted: 31 May 2021
Published: 1 June 2021

Publisher's Note: MDPI stays neutral with regard to jurisdictional claims in published maps and institutional affiliations.

In this issue, a range of new articles are presented, and we are fortunate to have a collection of empirical preclinical and human studies to assist in the development of understanding and progress in this area of research on improving health, and, in more detail, metabolic health. The studies in this Special Issue deal with various aspects of nutrition, as summarized below:

Energy Balance

Focused on the topic of energy balance, Cooney and colleagues report findings of a weight loss study in ageing Irish adults with overweight and adiposity-based chronic disease [1]. Participants had dietary energy requirements prescribed on the basis of either measured resting metabolic rate (mRMR) or estimated RMR by the prediction of Miffin [1]. A similar weight loss (>5%) over the short-term period of 12 weeks was seen in these two groups, together with a reduction in blood pressure, triglycerides, and glucose, thus reducing cardiovascular disease risk factors. Cumulatively, these data further support the use of RMR, either measured or estimated, to determine energy intake during a weight loss program [1].

Macronutrient Composition

In recreational athletes, Terink and colleagues elegantly showed, by using a cross-over study where athletes consumed one of two diets in random order with a wash-out period of >2 weeks in between, that a low-carbohydrate, high-fat (LCHF) diet resulted in reduced workload with metabolic effects and a pronounced exercise-induced cortisol response after 2 days, when compared to a high-carbohydrate (HC) diet. Although indications of adaptation were seen after 2 weeks on the LCHF diet, work output was still lower [2].

Specific Nutrients

Starting with the trace element iron, amongst others involved in oxidation–reduction reactions of energy metabolism, Rineau and colleagues focused on endurance capacity and fatigue, one of the main symptoms of iron deficiency [3]. They showed that iron deficiency without anemia in mice significantly reduced endurance and activity of the respiratory chain complex I in the predominantly slow-twitch musculus soleus, but not in the musculus quadriceps. This was seen without differences in complex IV activity in both muscles. They concluded that iron deficiency without anemia results in impaired mitochondrial complex I activity in skeletal muscles with predominantly oxidative metabolism, which might explain the observed reduction of fatigue and improved physical activity when correcting iron deficiency in humans [3].

In light of the increasing number of people with obesity and associated noncommunicable diseases nutritional approaches are highly warranted to combat developments of type 2 diabetes and the spectrum of conditions ranging from increased intrahepatic accumulation of triacylglycerols (fatty liver), hepatic steatosis, steatohepatitis (NASH) and end-stage liver disease,. Previously, it has been well reported that fish oils, and more specifically, the fatty acids eicosapentaenoic acid (EPA; 20:5n-3) and docosahexaenoic acid (DHA; 22:6n-3), contribute to health benefits, including, but not limited to, nonalcoholic fatty liver disease (NAFLD; reviewed by, e.g., Chang and colleagues (Prostaglandins Leukot. Essent. Fat. Acids, 2018). In this special issue, Sistilli and colleagues [4] and Bardova and colleagues [5] show some new insights revealing the nutritional power of these fatty acids as part of fish oil triglycerides or of krill oil (and its constituents), which includes high levels of phospholipids (PL) composed of a glycerol backbone with two fatty acids (either EPA or DHA) and a phosphate group modified with simple organic molecules such as choline, ethanolamine, or serine. Sistilli et al. showed impressive antisteatotic effects in the liver by krill oil versus fish oil using an obese, insulin-resistant mouse model of exacerbated NAFLD based on high-fat feeding at thermoneutral temperature. Moreover, effects were seen in both the prevention and reversal of hepatic steatosis. This was associated with improved hepatic insulin sensitivity and high plasma adiponectin levels [4].

Bardova et al., in contrast, investigated potential additive effects by combining nutritional and pharmacological interventions, using fish oil together with a first- or second-generation antidiabetic drug, thiazolidinedione (TZD). Focusing on white adipose tissue, increased fatty acid futile cycling (triacylglycerols → free fatty acids + glycerol → triacylglycerols) supporting energy dissipation was seen as an additive beneficial effect of fish oil and TZDs, together with increased metabolic health in these diet-induced obese mice. This included reduced body weight gain, and improvements in circulating and tissue metabolites and parameters of both lipid and glucose homeostasis [5].

Together, the studies of this Special Issue provide novel detailed insights into the physiological nature of the close relationship between (nutrients) our diet, energy metabolism, and physical functioning, and confirm the importance of this relationship for maintaining good health.

Author Contributions: A.G.N. and E.M.v.S. conceptualized and co-wrote this article. Both authors have read and agreed to the published version of the manuscript.

Funding: This research received no external funding.

Institutional Review Board Statement: Not applicable.

Informed Consent Statement: Not applicable.

Data Availability Statement: Not applicable.

Conflicts of Interest: The authors declare no conflict of interest.

References

1. Cooney, C.; Daly, E.; McDonagh, M.; Ryan, L. Evaluation of measured resting metabolic rate for dietary prescription in ageing adults with overweight and adiposity-based chronic disease. *Nutrients* **2021**, *13*, 1229. [CrossRef] [PubMed]
2. Terink, R.; Witkamp, R.F.; Hopman, M.T.E.; Siebelink, E.; Savelkoul, H.F.J.; Mensink, M. A 2 week cross-over intervention with a low carbohydrate, high fat diet compared to a high carbohydrate diet attenuates exercise-induced cortisol response, but not the reduction of exercise capacity, in recreational athletes. *Nutrients* **2021**, *13*, 157. [CrossRef] [PubMed]
3. Rineau, E.; Gueguen, N.; Procaccio, V.; Geneviève, F.; Reynier, P.; Henrion, D.; Lasocki, S. Iron deficiency without anemia decreases physical endurance and mitochondrial complex i activity of oxidative skeletal muscle in the mouse. *Nutrients* **2021**, *13*, 1056. [CrossRef] [PubMed]
4. Sistilli, G.; Kalendova, V.; Cajka, T.; Irodenko, I.; Bardova, K.; Oseeva, M.; Zacek, P.; Kroupova, P.; Horakova, O.; Lackner, K.; et al. Krill oil supplementation reduces exacerbated hepatic steatosis induced by thermoneutral housing in mice with diet-induced obesity. *Nutrients* **2021**, *13*, 437. [CrossRef] [PubMed]
5. Bardova, K.; Funda, J.; Pohl, R.; Cajka, T.; Hensler, M.; Kuda, O.; Janovska, P.; Adamcova, K.; Irodenko, I.; Lenkova, L.; et al. Additive effects of omega-3 fatty acids and thiazolidinediones in mice fed a high-fat diet: Triacylglycerol/fatty acid cycling in adipose tissue. *Nutrients* **2020**, *12*, 3737. [CrossRef] [PubMed]

Article

Evaluation of Measured Resting Metabolic Rate for Dietary Prescription in Ageing Adults with Overweight and Adiposity-Based Chronic Disease

Ciara Cooney , Ed Daly , Maria McDonagh and Lisa Ryan *

Department of Sport, Exercise and Nutrition, School of Science and Computing,
Galway-Mayo Institute of Technology, Galway Campus, Dublin Road, H91 T8NW Galway, Ireland;
ciara.cooney@gmit.ie (C.C.); ed.daly@gmit.ie (E.D.); Maria.McDonagh@gmit.ie (M.M.)
* Correspondence: lisa.ryan@gmit.ie; Tel.: +353-(0)91-742556

Abstract: The primary objective of this study was to compare weight changes in two groups of ageing Irish adults with overweight and adiposity-based chronic disease: participants who had dietary energy requirements prescribed on the base of measured RMR and participants whose RMR was estimated by a prediction equation. Fifty-four Caucasian adults (male $n = 25$; female $n = 29$, age 57.5 ± 6.3 years, weight 90.3 ± 15.1 kg, height 171.5 ± 9.5 cm, BMI 30.7 ± 4.6 kg/m^2) were randomly assigned to a dietary intervention with energy prescription based on either measured RMR or estimated RMR. RMR was measured by indirect calorimetry after an overnight fast and predicted values were determined by the Mifflin et al. (1990) prediction equation. All participants received individual nutritional counselling, motivational interviewing and educational material. Anthropometric variables, blood pressure, blood glucose and blood lipid profile were assessed over 12 weeks. Body weight at week 12 was significantly lower ($p < 0.05$) for both groups following dietary interventions, mRMR: -4.2%; eRMR: -3.2% of initial body weight. There was no significant difference in weight loss between groups. Overall, 20.8% mRMR and 17.4% of eRMR participants experienced clinically meaningful (i.e., $\geq 5\%$ of initial weight) weight reduction. Weight reduction in adults aged ≥ 50 years over the short term (12 weeks) favoured a reduction in blood pressure, triglycerides and glucose, thus reducing cardiovascular disease risk factors. This research indicates that employing a reduced-calorie diet using indirect calorimetry to determine energy needs when improving weight outcomes in adults (>50 years) with overweight and adiposity-based chronic disease is equal to employing a reduced-calorie diet based on the Mifflin et al. (1990) prediction equation. A reduced-energy diet based on mRMR or eRMR facilitates clinically meaningful weight reduction in adults (≥ 50 years) over the short term (12 weeks) and favours a reduction in blood pressure, triglycerides and glucose, thus reducing cardiovascular disease risk factors. Moreover, the addition of motivational interviewing and behaviour change techniques that support and encourage small behaviour changes is effective in short-term weight management.

Keywords: resting metabolic rate; prediction equation; ageing adults; overweight; obesity; adiposity-based chronic disease; energy balance; metabolism

Citation: Cooney, C.; Daly, E.; McDonagh, M.; Ryan, L. Evaluation of Measured Resting Metabolic Rate for Dietary Prescription in Ageing Adults with Overweight and Adiposity-Based Chronic Disease. *Nutrients* **2021**, *13*, 1229. https://doi.org/10.3390/nu13041229

Academic Editor: Arie Nieuwenhuizen and Evert van Schothorst

Received: 14 January 2021
Accepted: 6 April 2021
Published: 8 April 2021

Publisher's Note: MDPI stays neutral with regard to jurisdictional claims in published maps and institutional affiliations.

1. Introduction

The high prevalence of overweight (defined as a body mass index [BMI] ≥ 25 kg/m^2) and obesity (defined as a BMI ≥ 30 kg/m^2) among older Irish adults is a major health concern. Obesity is an 'adiposity-based chronic disease' (ABCD) that affects 35% of Irish adults aged 50 years and over, with a further 44% classified as overweight [1]. ABCD is associated with an increased risk of cardiovascular diseases (CVD), osteoarthritis, type 2 diabetes mellitus and impaired functionality [2,3]. Furthermore, older adults with ABCD are reported to be at greater risk of depression, disability and frailty than their age-matched counterparts of normal weight [1]. Central obesity, which affects over half (53%) of older

Irish adults, is characterised by increased abdominal adiposity and is associated with a greater risk of adverse metabolic and cardiovascular outcomes than overall obesity [4]. In addition, a higher prevalence of diabetes, high blood pressure and cardiac events was reported in older Irish adults with increased waist circumference (WC) and BMI than adults with normal WC and BMI [1].

Clinical guidelines available for the management and treatment of adults with obesity recommend lifestyle interventions involving diet, physical activity and behaviour modification for conventional obesity, with pharmacotherapy and surgical intervention for severe obesity cases [5–7]. Strategies to prevent weight gain, optimise weight loss and achieve long-term weight loss maintenance remain the hallmark of overweight and obesity treatment and management guidelines. Lifestyle weight management programmes consisting of reduced energy intake via calorie-restriction strategies and increased energy expenditure through increased physical activity are recommended with the support of a multidisciplinary team of health care professionals [6]. Calorie-restriction strategies such as low-calorie diets (LCD; 800–1600 kcal daily) may not be nutritionally complete and have long-term low compliance, and very low-calorie diets (VLCD < 800 kcal daily) require medical supervision due to the increased risk of medical complications [8]. For sustainable weight reduction the National Institute for Health and Care Excellence (NICE) recommend dietary approaches that reduce calories by 600 kcal/day, i.e., 600 kcal less than the individual requires to remain the same weight [9]. Similarly, a modest reduction in energy intake (500–750 kcal/day) is recommended for older adults by The American Society for Nutrition, the North American Association for the Study of Obesity (NAASO) and The Obesity Society [8]. In order to determine individual energy requirements, an assessment of resting metabolic rate (RMR) is recommended [10]. RMR is the main component of energy expenditure and accounts for up to 70% of total daily energy expenditure (TDEE) with the thermic effect of food and physical activity accounting for 10% and 20%, respectively. Energy expenditure associated with physical activity may be subdivided into energy utilised specifically for exercise, and non-exercise activity thermogenesis (NEAT) which involves maintaining posture and fidgeting [11]. RMR can be measured (mRMR) by respiratory indirect calorimetry (IC) or estimated by prediction equations (eRMR) [12]. IC is considered to be an accurate method of determining RMR [13]. IC is based on the indirect measure of the heat expended by nutrient oxidation, which is estimated by monitoring gas exchange, i.e., the volume of oxygen consumption (VO_2) and carbon dioxide (VCO_2) production over a period of time [13]. Such measures also provide information on energy substrate utilisation. The ratio of CO_2 production to O_2 consumption is known as the respiratory exchange ratio (RER) and represents fuel oxidation by IC [14]. During carbohydrate oxidation, there is an equal amount of CO_2 produced for O_2 consumed (RER = 1.0). During fat oxidation, there is less CO_2 produced for O_2 consumed [14]. A greater fractional oxidisation of fat (FAT-OX) as fuel is important for metabolic health, weight management, and body composition [15,16]. For instance, the skeletal muscle of adults with obesity, or insulin resistance displays an impaired ability to oxidise fat [17–20]. Decreases in skeletal muscle metabolic activity are associated with the ageing process and closely linked to age-related loss of muscle mass [20]. In addition, a high RER, which is indicative of low FAT-OX relative to carbohydrate oxidation, is predictive of both future body mass gain and fat mass (FM) regain after diet-induced reductions in body mass [21]. This information may be of particular relevance to ageing adults, particularly adults aged 50 years or older as a higher fat mass in relation to body mass accelerated the decline of muscle quality in this population [22]. Therefore, being able to accurately measure a person's ability to oxidise fat can have important implications for dietary manipulation strategies and may be more important in this age group.

IC has high reproducibility and is non-invasive; however, its use outside of clinical care settings is limited with commonly cited reasons including device expense, time required to carry out the measure, and the need for trained technicians to operate equipment and interpret test results [23]. Several metabolic rate prediction equations have been developed to calculate RMR and are frequently adopted by health care professionals to determine energy needs in order to develop nutritional support plans. The primary components used to develop prediction equations include weight, height, age, sex and body composition parameters [24]. Great variability has been reported in the accuracy of RMR prediction equations employed in adults with higher than normal BMIs, often resulting in the under- or over-estimation of an individual's specific calorie needs [25]. This may be because equations used were developed for a specific cohort such as normal weight individuals, whose characteristics differ from this population [26]. A minority of studies have validated prediction equations in adults with higher BMIs [27]. Moreover, the Mifflin et al. [28] prediction equation has been shown to provide a reliable estimate (78% within ± 10% limit of actual) of RMR in adults with normal weight and obesity [29,30]. Where IC is not available, the American Dietetic Association (ADA) recommend using the Mifflin et al. [28] prediction equation using actual body weight—males: $10 \times$ weight (kg) + $6.25 \times$ height (cm) $- 5 \times$ age (years) + 5; females: $10 \times$ weight (kg) + $6.25 \times$ height (cm) $- 5 \times$ age (years) $- 161$ to estimate RMR (eRMR) in adults with overweight or obesity [29].

A gradual decline in RMR and TDEE is associated with advancing age, diminished lean mass, energy restriction and weight loss [8,31–33]. Age-induced declines in RMR may be attributed to alterations of organ and tissue masses and diminished fat free mass (FFM) which accounts for the magnitude of resting metabolism [3,34]. Previous studies report decreases in RMR in response to negative energy balance and weight loss, with the observed decrease proportional to the energy deficit [35]. When the decline in RMR exceeds the magnitude predicted by the loss of body mass, metabolic adaptation occurs [36,37]. Metabolic adaptation can persist long term, resulting in implications for weight loss [38]. Furthermore, variation in physiological parameters such as: thyroid hormones, growth hormone, serum testosterone, leptin levels and sympathetic nervous system activity contribute to metabolic rate [39,40]. Given the various factors that influence RMR an accurate assessment is important for optimal dietary intake, with particular consideration to be given to the ageing process and associated disease states. Energy imbalance results in weight loss or weight gain and thus a major challenge in helping individuals reduce weight is to help individuals manage their energy balance. The main requirement of a dietary approach to weight reduction is that total energy intake must be less than energy expenditure [39]. Due to the decline in RMR with age (as mentioned above), it may be more important in this age group to measure RMR. Therefore, the aim of this study was to compare the efficacy of a dietary intervention (mRMR versus eRMR) on weight outcomes in ageing Irish adults (50 years and over) with overweight and obesity.

2. Materials and Methods

2.1. Study Design

This was a single-centre (GMIT) prospective study performed in a population of Irish adults classified as overweight or with obesity. Participants were randomly assigned to a 12-week dietary intervention, where energy intake was established using either (1) mRMR or (2) eRMR. Meal plans with prescribed energy intake and food options in line with habitual patterns were provided to each participant. The protocol consisted of six visits to the clinic which included a screening and familiarisation visit, a nutrition education visit and four measurement visits. Measurement visits were conducted at baseline, week 3, week 6 and week 12 of the dietary intervention. Anthropometric and RMR data were collected across all time points. At baseline and week 12 visits, participants provided capillary blood samples and had their blood pressure measured. The International Physical Activity Questionnaire Short Form (IPAQ-SF) was used to assess physical activity levels at baseline [41]. A 3-day food diary was used to determine habitual energy intake, eating patterns and food preferences at baseline and week 12. Diaries were assessed using Nutritics (Dublin, Ireland) professional dietary analysis software [42]. All measurements were conducted between 8:00 AM and 12:00 AM following an overnight fast and in a voided state. Participants were instructed to refrain from alcohol, nicotine and caffeine, and to avoid strenuous physical activity 10–12 h prior to the measurement visit. This study was conducted in accordance with the ethical principles expressed in the Declaration of Helsinki. Ethics approval was granted by the Research Ethics Committee of Galway Mayo Institute of Technology (GMIT), Ireland (RSC_AC230119). Written informed consent was provided by all participants prior to their inclusion in this study.

2.2. Sample Size Calculation

Power analyses were performed prior to the start of this study in order to identify an appropriate sample size. Based on data reported in the literature, 23 individuals were required per group (46 in total) to detect a 5% loss in body weight at a significance level of 0.05 and power of 80%. To allow for a participant withdrawal rate of 20%, the recruitment target was set at 56 participants.

2.3. Participants

Fifty-six adults (male $n = 26$; female $n = 30$) were recruited via advertisements in local community centres, libraries, general practitioners and health care centres to take part in this study. Inclusion criteria were adults aged 50 years or greater with a BMI greater than or equal to 25 kg/m^2. Participants were excluded from this study if they had any health conditions or were taking medication known to influence the measurement of RMR or body composition, experienced weight loss of 5% or greater in the previous 3 months, a past or present history of eating disorders or disorders that would be incompatible with safe and successful participation in this study, as determined by the investigators.

2.4. Randomisation and Allocation

Participants were allocated by computer-generated randomisation to one of two intervention groups (1) mRMR or (2) eRMR. The group assignment was stratified using a computerised program (Excel) to ensure equal distribution of BMI in the two groups. The random allocation of the intervention groups was carried out by a separate investigator who was not involved with data collection (LR).

2.5. Treatment Protocol

Participants in the mRMR group received a meal plan with energy prescription based on mRMR using a portable IC (ECAL Energy Testing Solutions, UK). The ECAL is a validated open-circuit portable calorimeter that provides practitioners and users with information concerning energy metabolism such as resting energy expenditure and RER. The device utilises breath-by-breath measurement of gas exchange through a plastic mouthpiece and tubing for gas collection. The VO_2 and VCO_2 are measured using a small mixing chamber. VO_2 is measured using a galvanic fuel cell oxygen analyser. VCO_2 is measured using a patented ultra-low power VCO_2 analyser which uses Light Emitting Diode (LED) and detector technology in a novel non-dispersive near-infrared absorption sensor. The meal plan consisted of a 7-day menu. Individual energy requirements were calculated from mRMR and a physical activity factor corresponding to a category (low = 1.2, moderate = 1.55, high = 1.725) was applied to participants RMR to account for individual activity requirements. A subsequent energy deficit of 500 kcal was prescribed to promote a 0.45 kg per week weight reduction (mRMR $\times$ PAL $-$ 500). Energy requirements were adjusted accordingly following repeated measures of RMR and participants received meal plans to reflect changes required in energy needs. Energy information such as RER was used to advise on dietary modification. Participants with a RER $\geq$ 0.75 (less than 16% fat burning efficiency) were advised to modify carbohydrate intake. Participants with optimal fat oxidation of >80% were advised to continue to follow healthy eating guidelines while maintaining prescribed energy intake.

Participants in the eRMR group received a meal plan with energy prescription based on eRMR using the Mifflin et al. [28] prediction equation. The meal plan consisted of a 7-day menu following national healthy eating guidelines [43]. The Mifflin et al. [28]

prediction equation (males $10 \times$ weight (kg) + $6.25 \times$ height (cm) $- 5 \times$ age (years) + 5; females $10 \times$ weight (kg) + $6.25 \times$ height (cm) $- 5 \times$ age (years) $- 161$) was used to inform energy prescription. The eRMR was multiplied by a physical activity level (PAL) factor and a subsequent 500 kcal was subtracted from daily energy requirements to promote a 0.45 kg per week weight reduction (eRMR $\times$ PAL $-$ 500).

All participants attended a registered dietician led nutritional educational presentation (1.5 h) at GMIT. Core topics included healthy eating guidelines, weight management, physical activity and common age-related nutrition issues such as constipation, bloating, irritable bowel symptoms and CVD. Physical activity recommendations were provided as per national guidelines, HSE [44].

2.6. Compliance

Compliance to the prescribed dietary intervention was monitored from changes in body weight reviewed at week 3 and week 6. Individual consultations of 30 min were carried out at follow-up visits to encourage compliance. The aim of the consultation was to listen to the participant, identify barriers that may be contributing to challenges faced and guide the participant to potential solutions using behaviour change skills and motivational interviewing techniques. Visual aids such as the Irish Food Pyramid, Eat Well plate, disposable cups and food labels were used to encourage the adoption of the dietary guidelines and to describe portion sizes and food choices to the participants. Further individual support was provided via e-mail as required.

2.7. Measures

Body weight and anthropometric measurements were assessed at baseline, week 3, week 6 and week 12. Measurements were taken with participants in a fasted and voided state, wearing light clothing and shoes and socks removed. Height was recorded to the nearest 0.5 cm using a stadiometer (Seca Ltd., Birmingham, UK). Body weight and body composition (percentage fat mass and FFM) were measured using bioelectrical impedance analyser (BIA) Tanita BC-418 MA (Tanita UK Limited, Yiewsley, UK). BMI was calculated as weight in kilograms divided by height in metres squared (kg/m^2). Waist circumference (WC) and hip circumference (HC) were assessed using steel tape (Lufkin W606PM) according to the International Society for the Advancement of Kinanthropometry (ISAK) standards for anthropometric assessment [45].

Resting blood pressure was assessed at baseline and week 12 using an automated sphygmomanometer (Omron M500 HEM-7321-D, Milton Keynes, UK) with participants in a seated position. Three measurements were taken with the average recorded. Blood samples were collected at baseline and week 12 of the dietary intervention. Blood samples were obtained in a fasted state by standard laboratory techniques (finger-stick procedure). Capillary whole blood was tested for fasting blood glucose (mmol/L) using the Accutrend Blood Glucose Monitor (Roche Diagnostics, Dublin, Ireland). Total cholesterol (TC) (mmol/L), triglycerides (mmol/L), HDL (mmol/L), calculated LDL and TC:HDL were assessed using Cardio-Chek monitor (Roche Diagnostics, Dublin, Ireland).

RMR and RER were measured throughout the study period. Participants were instructed to refrain from alcohol and caffeine, and to avoid strenuous physical activity 10–12 h prior to each measurement visit. Participants arrived between 08:00 and 13:00 h following an overnight fast (10–12 h before testing time). Prior to each test, the ECAL calorimeter was calibrated as per manufacturer's instructions. Following a rest period of 10 min, participants lay in a semi-reclined, comfortable position in a quiet room and were reminded to stay awake. A mouthpiece and nose clip were employed, and participants were instructed to breathe in and out through the mouthpiece as normal. Measurements were recorded for 10 min. Upon completion of the test, the mouthpiece and nose clip were removed.

2.8. Statistical Analysis

Analysis were performed using Statistical Package for the Social Sciences (SPSS) for Windows (version 25.0; IBM Corporation, Armonk, NY, USA). Normality of data and outliers were assessed using Shapiro–Wilk and boxplot, respectively. Homogeneity of variances and covariances were assessed by Levene's test and Box's M test, respectively. Independent-samples *t*-tests were used to determine differences between the groups at baseline for normally distributed continuous variables. Mann–Whitney U tests were used to assess differences between the groups at baseline for non-normally distributed variables and/or variables with identified outliers. Parametric variables are presented as the mean ± standard deviation (SD) and non-parametric variables as the median (Md) (interquartile range [IQR]). Two-way mixed analysis of variance (ANOVA) assessed the level of difference between groups and within groups overtime using completer analysis for each respective dependent variable. All post hoc tests were carried out with Bonferroni corrections. When sphericity was violated Greenhouse–Geisser correction was reported. Outliers confirmed as genuine data points were included in the analysis. Where data were not normally distributed, the two-way mixed ANOVA was run regardless and reported, as ANOVAs are considered fairly robust to deviations from normality. The level of significance was accepted at $p < 0.05$.

3. Results

3.1. Participants

Participant and study flow are illustrated in Supplementary Material Figure S1. From March to April 2019, fifty-six adults were recruited to take part in this study. Two participants (male $n = 1$, female $n = 1$) withdrew prior to baseline testing, citing time commitments conflicting with the study requirements, resulting in fifty-four Caucasian adults (male $n = 25$; female $n = 29$) with a mean ± SD height, body mass, age and BMI of 171.5 ± 9.5 cm, 90.3 ± 15.1 kg, 57.5 ± 6.3 years, and 30.7 ± 4.6 kg/m^2, respectively, at baseline. Baseline participant characteristics and baseline measures are presented in Table 1. Anthropometric measurements (weight, BMI, WC, HC, WHR, body fat percent and muscle mass) and clinical outcomes (blood pressure) across the intervention period are presented in Table 2. Primary outcome pre and post weight are presented in Table 3. Weight, assessed at baseline, week 3, week 6 and week 12 is illustrated in Figure 1. Individual response to percent weight change is illustrated in Figure 2. Biochemical outcomes at baseline and week 12 are presented in Table 4. Changes in biochemical markers (glucose, TC, HDL, HDL:TC, LDL, triglycerides) are illustrated in Figures 3–8, respectively. Metabolic outcomes across the intervention period and estimated energy intake are presented in Table 5. Male and female energy intake at baseline and week 12 and prescribed energy are presented in Table 6. Prescribed energy intake versus reported energy intake from a 3-day food diary is presented in Table 7.

3.2. Weight Change

There were no significant differences in baseline weight for mRMR (Md = 88.50, $n = 29$) and eRMR participants (Md = 92.90, $n = 25$), U = 420, z = 0.998, $p = 0.32$, r = 0.14 (Table 2). There was no significant interaction between the intervention groups and time for weight, F(1.478, 54.677) = 0.57, $p = 0.518$, partial $\eta^2 = 0.02$. There was no significant main effect of group on the mean weight F(1, 37) = 0.789, $p = 0.380$, partial $\eta^2 = 0.021$. There was a significant main effect for time on the difference in mean weight at the different time points, F(1.478, 54.677) = 26.726, $p < 0.0005$, partial $\eta^2 = 0.419$. Post hoc analysis revealed that body weight was significantly lower (all $< p = 0.005$) at week 3 (1.9%), week 6 (3.0%) and week 12 (3.3%) compared to baseline for both groups. Weight at week 6 (1.2%) and week 12 (1.4%) was significantly (both $< p = 0.003$) lower than weight at week 3 for both groups. Post hoc analysis revealed no significant difference between weight at week 6 versus week 12 (0.1%, $p = 1.0$) for both groups (Figure 1).

Table 1. Baseline participant characteristics.

	n	mRMR Group	n	eRMR Group	p Value
Sex males (M) females (F)	M10/F19		M15/F10		
Age (years)	29	56.7 ± 5.3	25	58.6 ± 7.1	
Height (cm)	29	170.3 ± 9.5	25	173.1 ± 8.8	
Weight (kg)	29	88.5 (81.0, 94.8)	25	92.9 (81.0, 98.9)	† p = 0.32
BMI (kg/m^2)	29	29.3 (26.8, 33.4)	25	29.5 (27.7, 32.6)	† p = 0.23
WC (cm)	28	106.0 (99.3, 115.3)	25	112.5 (102.5, 116.0)	† p = 0.67
HC (cm)	28	112.3 (106.0, 126.1)	25	113.0 (107.3, 123.0)	† p = 0.98
WHR	28	0.9 ± 0.1	25	1.0 ± 0.1	φ p = 0.15
Body Fat (%)	29	37.5 ± 8.6	25	35.7 ± 7.5	† p = 0.42
Muscle Mass (kg)	29	49.1 (45.1, 63.0)	25	59.6 (44.6, 67.3)	φ p = 0.22
BPsys (mmHg)	29	126.0 (115.5, 136.0)	25	135.0 (124.0, 151.0)	† p = 0.04
BPdia (mmHg)	29	83.0 (74.5, 87.5)	25	87.0 (78.0, 92.0)	† p = 0.10,
Glucose (mmol/L)	28	4.9 (4.5, 5.5)	25	5.3 (4.5, 5.8)	† p = 0.45
TC (mmol/L)	28	4.7 (4.1, 5.3)	22	3.9 (3.5, 4.5)	† p = 0.02
HDL (mmol/L)	28	1.3 (1.1, 1.8)	23	1.2 (1.1, 1.6)	† p = 0.76
TC:HDL Ratio	28	3.4 (2.6, 4.1)	21	3.0 (2.4, 3.4)	† p = 0.08
LDL (mmol/L)	24	2.8 (2.5, 3.1)	19	2.2 (1.7, 2.6)	† p = 0.00
TG (mmol/L)	25	1.1 (0.9, 1.4)	19	1.2 (1.0, 1.6)	† p = 0.34
mRMR (kcal)	29	1604.0 (1374.0, 2011.5)	23	1691.0 (1455.0, 2067.0)	† p = 0.80
RER	29	0.8 (0.7, 0.9)	23	0.8 (0.7, 0.9)	† p = 0.38
eRMR kcal	29	1560.3 ± 221.7	24	1639.3 ± 272.2	φ p = 0.25
Energy Intake (kcal)	21	2195.0 (1863.5, 2755.0)	19	2129.0 (1880.0, 2586.0)	† p = 0.68

Values are presented as the mean ± SD or median (25th–75th percentile), if data were non-parametric. n = number of participants with data available for each outcome. IQR, interquartile range; BMI, body mass index; WC, waist circumference; HC, hip circumference; WHR, waist to hip ratio; BPsys, systolic blood pressure; BPdia, diastolic blood pressure; TC, total cholesterol; HDL, high-density lipoprotein; TC:HDL, total cholesterol to high-density lipoprotein ratio; LDL, low-density lipoprotein; TG, triglycerides; RMR, resting metabolic rate; RER, respiratory exchange ratio; φ denotes independent-samples *t*-test; † denotes Mann–Whitney U test.

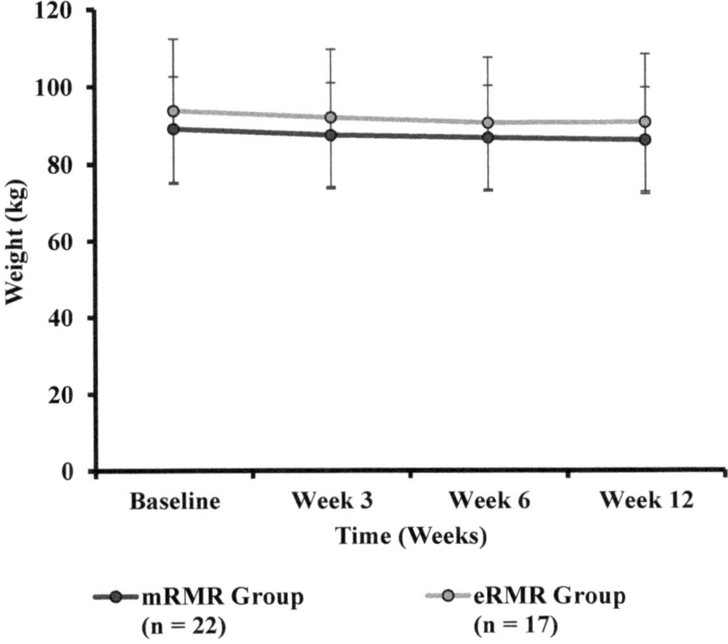

Figure 1. Weight (kg) across the intervention period for the mRMR (n = 22) and eRMR (n = 17) groups.

Table 2. Anthropometric and clinical outcomes across the intervention period.

		mRMR Group					eRMR Group						
	n	Baseline	Week 3	Week 6	Week 12	n	Baseline	Week 3	Week 6	Week 12	Time Effect, p	Group Effect, p	Time-by-Group Interaction, p
Weight (kg)	22	89.0 ± 13.6	87.3 ± 13.7	86.7 ± 13.6	86.0 ± 13.7	17	93.7 ± 18.7	92.0 ± 17.8	90.5 ± 17.0	90.7 ± 17.8	$p < 0.0005$ *	$p = 0.38$	$p = 0.52$
BMI (kg/m^2)	24	30.4 ± 5.2			29.4 ± 5.2	23	31.1 ± 4.4			30.4 ± 4.0	$p < 0.0005$ *	$p = 0.53$	$p = 0.57$
WC (cm)	23	109.0 ± 13.9			98.2 ± 11.8	23	113.2 ± 12.3			105.5 ± 11.1	$p < 0.0005$ *	$p = 0.10$	$p = 0.22$
HC (cm)	23	117.2 ± 14.5			108.6 ± 10.4	23	117.0 ± 13.7			108.8 ± 8.1	$p < 0.0005$ *	$p = 0.99$	$p = 0.92$
WHR	23	0.9 ± 0.1			0.9 ± 0.1	23	1.0 ± 0.1			1.0 ± 0.1	$p = 0.05$ *	$p = 0.02$ *	$p = 0.12$
Body Fat (%)	22	36.5 ± 9.2	36.7 ± 9.6	35.3 ± 9.5	36.9 ± 11.3	17	36.3 ± 7.0	35.3 ± 7.7	35.0 ± 7.2	35.6 ± 7.7			
Muscle Mass (kg)	22	54.0 ± 11.7	52.7 ± 11.8	53.4 ± 11.4	51.8 ± 12.8	17	57.2 ± 13.3	56.9 ± 12.9	56.5 ± 13.3	56.0 ± 13.2	$p < 0.0005$ *	$p = 0.36$	$p = 0.37$
BPsys (mmHg)	23	124.0 ± 15.4			121.2 ± 14.3	23	137.8 ± 20.5			130.8 ± 16.2	$p < 0.0005$ *	$p = 0.01$ *	$p = 0.30$
BPdia (mmHg)	23	81.2 ± 11.7			77.9 ± 9.0	23	86.7 ± 9.8			83.8 ± 9.7	$p < 0.0005$ *	$p = 0.04$	$p = 0.88$

Values are presented as the mean ± SD. n = number of participants with complete data available for each outcome. p value obtained from a two-way mixed ANOVA test. * denotes significant difference, $p < 0.05$. BMI, body mass index; WC, waist circumference; HC, hip circumference; WHR, waist to hip ratio; BPsys, systolic blood pressure; BPdia, diastolic blood pressure.

Table 3. Primary outcome weight pre and post intervention.

		mRMR Group			eRMR Group				
	n	Baseline	Week 12	n	Baseline	Week 12	Time Effect, p	Group Effect, p	Time-by-Group Interaction, p
Weight (kg)	24	88.0 ± 13.7	85.2 ± 13.6	23	93.5 ± 17.4	91.4 ± 16.8	$p < 0.0005$ *	$p = 0.20$	$p = 0.56$

Values are presented as the mean ± SD. n = number of participants. p value obtained from a two-way mixed ANOVA test on pre and post data only. *denotes significant difference from baseline, $p < 0.05$.

Table 4. Biochemical outcomes at baseline and week 12.

		mRMR Group			eRMR Group				
	n	Baseline	Week 12	n	Baseline	Week 12	Time Effect, p	Group Effect, p	Time-by-Group Interaction, p
Glucose (mmol/L)	23	5.0 ± 0.7	4.7 ± 0.7	22	5.4 ± 0.9	5.1 ± 0.8	$p < 0.0005$ *	$p = 0.08$	$p = 0.74$
TC (mmol/L)	23	4.5 ± 1.1	4.4 ± 1.0	20	4.1 ± 0.7	4.2 ± 0.9	$p = 0.92$	$p = 0.18$	$p = 0.45$
HDL (mmol/L)	23	1.4 ± 0.4	1.3 ± 0.4	21	1.4 ± 0.4	1.3 ± 0.3	$p < 0.0005$ *	$p = 0.91$	$p = 0.15$
TC:HDL Ratio	23	3.4 ± 0.9	3.6 ± 0.9	19	3.0 ± 0.6	3.2 ± 0.8	$p = 0.09$	$p = 0.07$	$p = 0.87$
LDL (mmol/L)	19	2.7 ± 0.7	2.6 ± 0.8	17	2.0 ± 0.6	2.3 ± 0.8	$p = 0.43$	$p = 0.02$ *	$p = 0.28$
TG (mmol/L)	20	1.2 ± 0.5	1.1 ± 0.4	18	1.3 ± 0.4	1.2 ± 0.4	$p = 0.16$	$p = 0.63$	$p = 0.67$

Values are presented as the mean ± SD. n = number of participants with complete data available for each outcome. p value obtained from a two-way mixed ANOVA test. TC, total cholesterol; HDL, high-density lipoprotein; TC:HDL, total cholesterol to high-density lipoprotein ratio; LDL, low-density lipoprotein; TG, triglycerides. *denotes significant difference, $p < 0.05$.

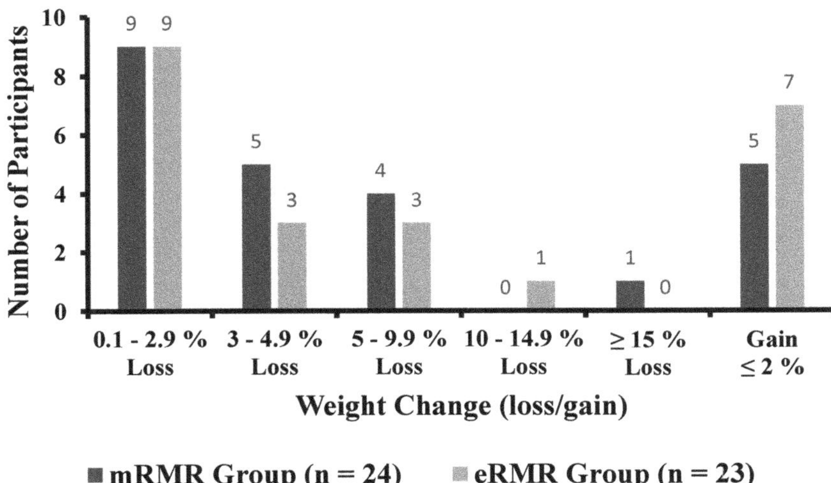

Figure 2. Individual response to weight change for participants completing pre and post measures in the mRMR (*n* = 24) and eRMR groups (*n* = 23).

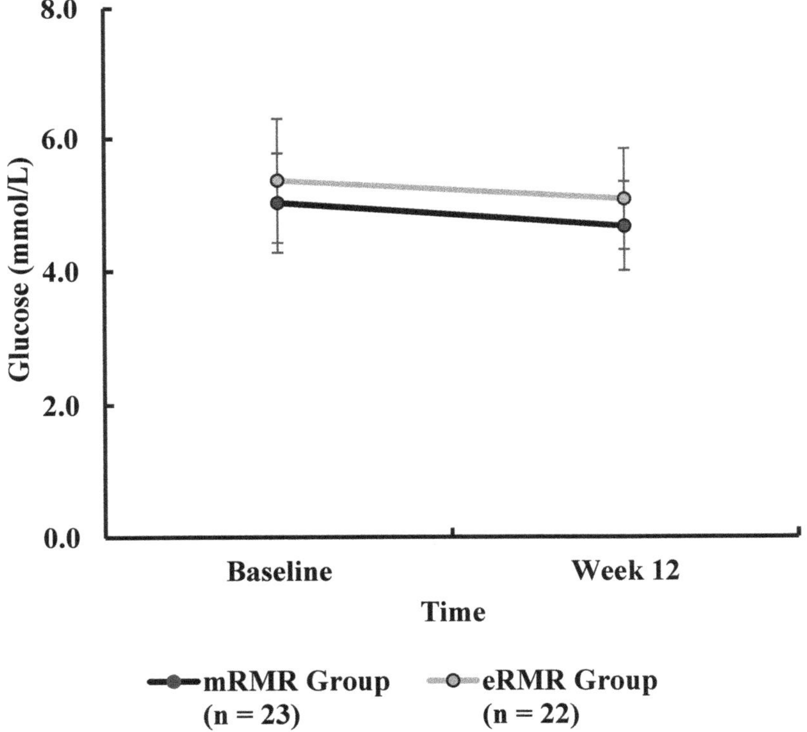

Figure 3. Blood glucose (mmol/L) measured at baseline and week 12 for the mRMR (*n* = 23) and eRMR groups (*n* = 22).

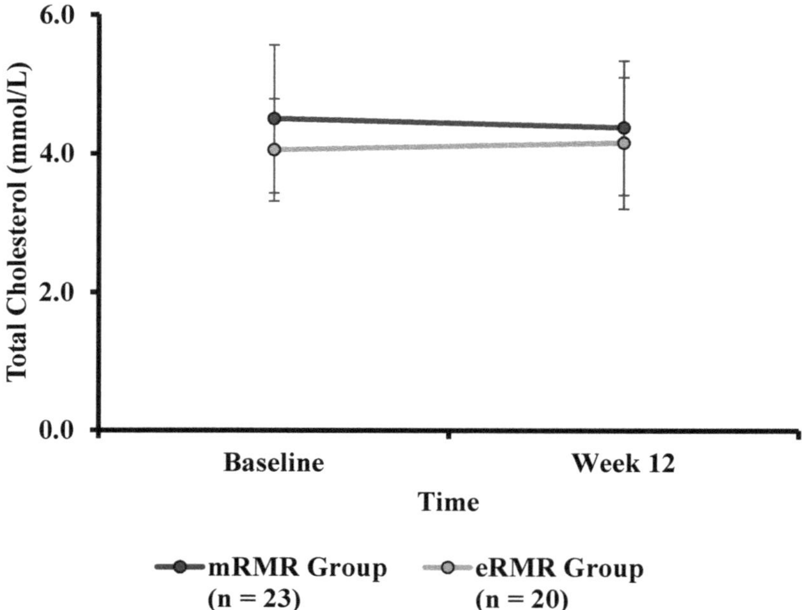

Figure 4. Total cholesterol (mmol/L) measured at baseline and week 12 for the mRMR (n = 23) and eRMR groups (n = 20).

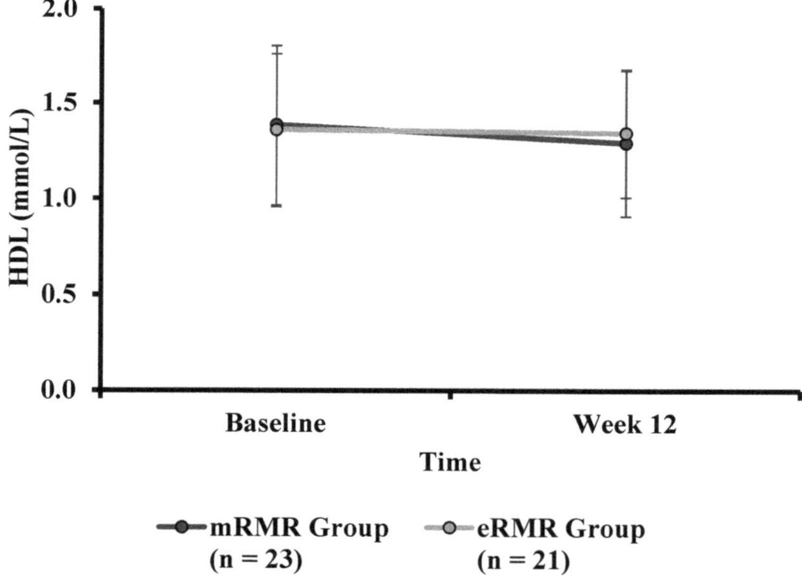

Figure 5. High-density lipoprotein (mmol/L) measured at baseline and week 12 for the mRMR (n = 23) and eRMR groups (n = 20).

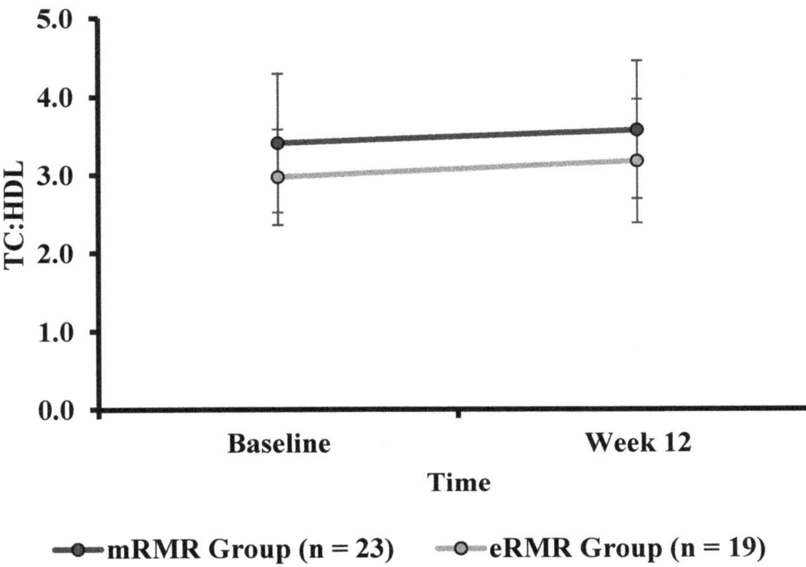

Figure 6. Total cholesterol to high-density lipoprotein ratio calculated at baseline and week 12 in both the mRMR (*n* = 23) and estimated eRMR groups (*n* = 19) groups.

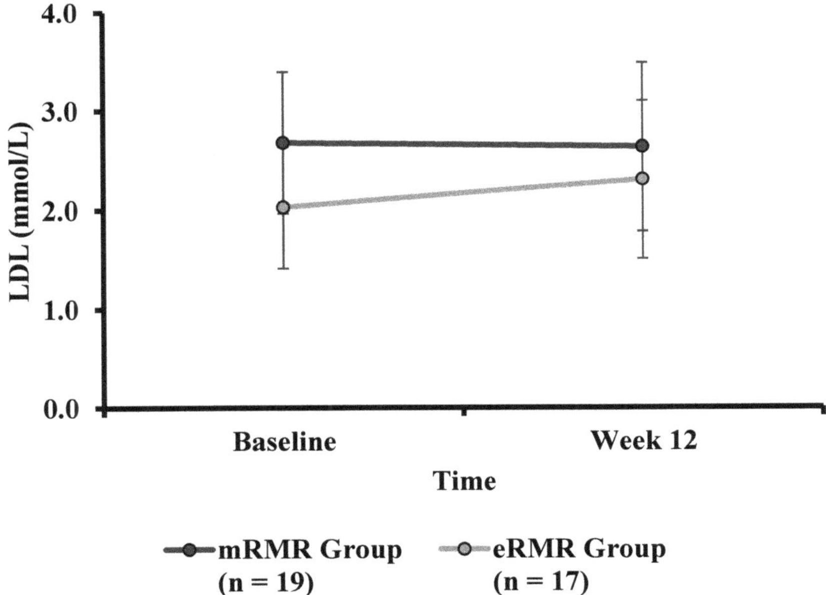

Figure 7. Low-density lipoprotein (mmol/L) calculated at baseline and week 12 for the mRMR (*n* = 19) and eRMR groups (*n* = 17).

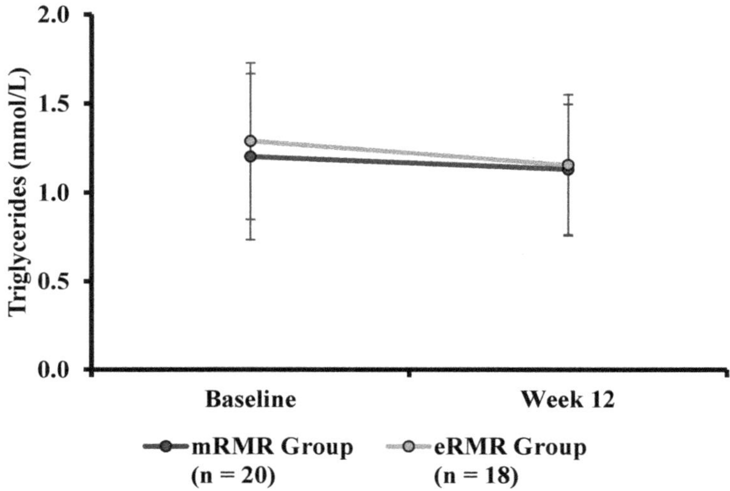

Figure 8. Triglycerides (mmol/L) measured at baseline and week 12 for the mRMR (n = 20) and eRMR groups (n = 18).

Table 5. Metabolic outcomes and estimated energy intake for groups across the intervention period.

Variable		mRMR Group					eRMR Group				Time Effect, p	Group Effect, p	Time-by-Group Interaction, p
	n	Baseline	Week 3	Week 6	Week 12	n	Baseline	Week 3	Week 6	Week 12			
mRMR (kcal)	22	1764.3 ± 547.6	1687.7 ± 505.8	1737.8 ± 475.1	1688.3 ± 454.6	16	1813.8 ± 410.4	1688.8 ± 469.6	1787.4 ± 566.5	1780.5 ± 518.8	$p = 0.22$	$p = 0.38$	$p = 0.75$
RER	22	0.8 ± 0.2	0.8 ± 0.1	0.8 ± 0.1	0.8 ± 0.1	16	0.8 ± 0.1	0.8 ± 0.1	0.8 ± 0.1	0.8 ± 0.1	$p = 0.32$	$p = 0.75$	$p = 0.34$
eRMR (kcal)	22	1578.5 ± 227.7	1576.0 ± 230.9	1553.7 ± 222.5	1547.0 ± 225.0	17	1656.9 ± 297.7	1648.4 ± 276.4	1624.1 ± 282.2	1625.5 ± 288.0	$p < 0.0005$ *	$p = 0.37$	$p = 0.79$
eEI (kcal)	15	2327.3 ± 827.1			1841.1 ± 534.3	15	2117.3 ± 562.8			1645.1 ± 433.3	$p < 0.0005$ *	$p = 0.28$	$p = 0.96$

Values are presented as the mean ± SD. n = number of participants with complete data set available for each variable. p value obtained from a two-way mixed ANOVA test. * denotes significant difference, $p < 0.05$. mRMR, measured resting metabolic rate; RER, respiratory exchange ratio; eRMR, estimated resting metabolic rate (Mifflin-St. Jeor equation); eEI, estimated energy intake (group analysis from a 3-day food diary).

Table 6. Estimated energy intake at baseline and week 12 for male and female participants.

	Intervention Group	Sex	n	Baseline	Week 12	Sex	n	Baseline	Week 12
eEI (kcal)	mRMR	M	5	2462 ± 973	2052 ± 597	F	10	2260 ± 793	1736 ± 473
	eRMR	M	8	2267 ± 724	1804 ± 445	F	15	1947 ± 253	1463 ± 367

Values are presented as the mean ± SD. n = number of participants with complete data available at each time point. eEI, estimated energy intake (from a a3 day food diary) for male and female participants.

Table 7. Prescribed energy intake versus estimated energy intake at week 12 for male and female participants.

	Intervention Group	Sex	n	Prescribed	Estimated Week 12	Sex	n	Prescribed	Estimated Week 12
Energy Intake (kcal)	mRMR	M	5	2180 ± 253	2052 ± 597	F	10	1584 ± 271	1736 ± 473
	eRMR	M	8	2100 ± 236	1804 ± 445	F	15	1500 ± 141	1463 ± 367

Values are presented as the mean ± standard deviation. n = number of participants with complete data set available at each time point. Estimated 12-week daily average energy intake from a 3-day food diaries.

3.3. Individual Response to Weight

A total of 37.5% of participants in the mRMR group and 39.1% of participants in the eRMR group reduced between 0.1 and 2.9% of initial body weight. A weight reduction between 3 and 4.9% was observed in 20.8% and 13% of participants in the mRMR and eRMR groups, respectively. A 5–9.9% reduction from initial body weight was observed in 16.7% and 13% of participants in the mRMR and eRMR groups, respectively. Of the participants in the eRMR, 4.3% experienced a 10–14.9% weight reduction. Of the participants in the mRMR 4.2% experienced a weight reduction of $\geq$15%. Overall, 20.8% mRMR and 17.4% of eRMR participants experienced clinically meaningful (i.e., $\geq$5% of initial weight) weight reduction. Weight gain of $\leq$2% was observed in 20.8% and 30.4% of participants in the mRMR and eRMR groups, respectively (Figure 2).

3.4. Body Mass Index

There were no significant differences between groups for baseline BMI (Table 1) of mRMR (Md = 106.00, n = 28) and eRMR (Md = 112.50, n = 25), U = 417, z = 1.194, p = 0.232, r = 0.16. There was no significant interaction between the intervention and time on BMI (Table 2), $F(1, 45)$ = 0.335, p = 0.566, partial η^2 = 0.007. There was no significant effect for group, when comparing BMI, $F(1, 45)$ = 0.394, p = 0.534, partial η^2 = 0.009. The main effect of time showed a statistically significant difference in BMI, across time points, $F(1, 45)$ = 25.801, p < 0.0005, partial η^2 = 0.364. Post hoc analysis revealed that BMI was significantly reduced (p = 0.001) at week 12 (29.9 kg/m^2, SE = 0.7) compared to baseline (30.74 kg/m^2, SE = 7.0).

3.5. Waist Circumference

There were no significant differences for WC at baseline (Table 1) between mRMR (Md = 29.30, n = 29) and eRMR participants (Md = 29.46, n = 25), U = 387, z = 0.43, p = 0.67, r = 0.06. There was no significant interaction (intervention x time) for WC (Table 2), $F(1, 44)$ = 1.57, p = 0.22, partial η^2 = 0.03. The main effect of group showed that there was no statistically significant difference in WC between the intervention groups, $F(1, 44)$ = 2.77, p = 0.10, partial η^2 = 0.06. The main effect of time showed a statistically significant difference in WC over time, $F(1, 44)$ = 58.083, p < 0.0005, partial η^2 = 0.569. Post hoc analysis revealed that WC was significantly reduced (p = 0.001) at week 12 (-9.3 cm) compared to baseline (111.1 cm SE = 1.9).

3.6. Hip Circumference

There was no significant difference in HC at baseline (Table 1) between mRMR (Md = 112.25, n = 28) and eRMR participants (Md = 113.00, n = 25), U = 351, z = 0.027, p = 0.979, r = 0.00. There was no significant interaction between the intervention and time on HC (Table 2), $F(1, 44)$ = 0.010, p = 0.921, partial η^2 = 0.000. The main effect of group showed that there were no statistically significant differences in HC between the intervention groups (Table 2), $F(1, 44)$ = 0.000, p = 0.99, partial η^2 = 0.000. The main effect of time showed a substantial statistically significant difference in HC over time (Table 2), $F(1, 44)$ = 20.586, p < 0.0005, partial η^2 = 0.32. Post hoc analysis revealed that HC was significantly reduced (p = 0.001) at week 12 (-8.4 cm) compared to baseline (117.1 cm, SE = 2.1).

3.7. Waist to Hip Ratio

There were no significant differences in WHR between mRMR (0.93 $\pm$ 0.077) and eRMR groups (0.96 $\pm$ 0.082; t(51) = -1.479, p = 0.145) at baseline (Table 1). There was no significant interaction between the intervention and time on WHR (Table 2), $F(1, 44)$ = 2.528, p = 0.119, partial η^2 = 0.054. The main effect of group showed a substantial significant difference in WHR between the intervention groups, $F(1, 44)$ = 5.751, p = 0.02, partial η^2 = 0.12. The main effect of time showed a substantial statistically significant difference in WHR at the different time points, $F(1, 44)$ = 4.085, p = 0.049, partial η^2 = 0.09. Post hoc analysis revealed that WHR was significantly reduced ($p = 0.049$) at week 12 (0.9 SE = 0.01) compared to baseline (1.0 SE = 0.01).

3.8. Percent Body Fat

There were no significant differences in baseline body fat percent (Table 1) between mRMR group (37.53 ± 8.57) and eRMR groups (35.72 ± 7.53; $t(52) = 0.818$, $p = 0.417$). There was no significant interaction between the intervention groups and time on body fat (Table 2), $F(1.785, 66.044) = 0.705$, $p = 0.482$, partial $\eta^2 = 0.019$. The main effect of group showed that there was no significant difference in body fat between the intervention groups, $F(1, 37) = 0.084$, $p = 0.773$, partial $\eta^2 = 0.002$. The main effect of time showed no significant difference in body fat over time, $F(1.785, 66.044) = 2.259$, $p = 0.118$, partial $\eta^2 = 0.058$.

3.9. Muscle Mass

There were no significant differences in baseline muscle mass (Table 1) between mRMR (Md = 49.10, $n = 29$) and eRMR participants (Md = 59.60, $n = 25$), U = 433, $z = 1.22$, $p = 0.221$, r = 0.17. There was no significant interaction between the intervention groups and time on muscle mass (Table 2), $F(1.980, 73.275) = 1.017$, $p = 0.366$, partial $\eta^2 = 0.027$. The main effect of group showed that there was no significant difference in muscle mass between the intervention groups, $F(1, 37) = 0.846$, $p = 0.36$, partial $\eta^2 = 0.02$. The main effect of time showed a substantial significant difference in muscle mass over time, $F(1.980, 73.275) = 6.227$, $p < 0.0005$, partial $\eta^2 = 0.14$. Post hoc analysis revealed that muscle mass was significantly reduced (both $< p = 0.035$) at week 3 (0.8 kg) and week 12 (2.1 kg) compared to baseline (55.6 kg, SE = 2.0). Post hoc analysis revealed no significant difference between muscles mass at week 6 and week 12 versus week 3 (both $> p = 0.65$). Post hoc analysis revealed no significant difference between muscle mass at week 6 versus week 12 (0.1%, $p= 0.21$).

3.10. Systolic Blood Pressure

There was a significant difference in systolic blood pressure at baseline (Table 1) between mRMR (Md = 126.00, $n = 29$) and eRMR participants (Md = 135.00, $n = 25$), U = 481, $z = 2.065$, $p = 0.039$, r = 0.28. There was no significant interaction between the intervention and time on systolic blood pressure (Table 2), $F(1, 44) = 1.124$, $p = 0.295$, partial $\eta^2 = 0.025$. The main effect of group showed a substantial significant difference in systolic blood pressure between the intervention groups, $F(1, 44) = 6.654$, $p = 0.013$, partial $\eta^2 = 0.13$. Post hoc analysis revealed systolic blood pressure was significantly lower in the mRMR group when compared to the eRMR group (-11.71 mmHg, $p = 0.01$). The main effect of time showed a significant difference in systolic blood pressure over time, $F(1, 44) = 6.305$, $p < 0.0005$, partial $\eta^2 = 0.13$. Post hoc analysis revealed that systolic blood pressure was significantly reduced at week 12 (-4.9 mmHg, $p = 0.02$) compared to baseline (130 mmHg, SE = 2.7).

3.11. Diastolic Blood Pressure

There were no significant differences in baseline diastolic blood pressure (Table 1) between mRMR (Md = 83.00, $n = 29$) and eRMR participants (Md = 87.00, $n = 25$), U = 457, $z = 1.65$, $p = 0.099$, r = 0.22. There was no significant interaction between the intervention and time on diastolic blood pressure (Table 2), $F(1, 44) = 0.022$, $p = 0.884$, partial $\eta^2 = 0.000$. The main effect of group showed a substantial statistically significant difference in diastolic blood pressure between the intervention groups, $F(1, 44) = 4.524$, $p = 0.039$, partial $\eta^2 = 0.093$. Post hoc analysis revealed that diastolic blood pressure was significantly lower in the eRMR group when compared to mRMR (-5.7 mmHg, $p = 0.04$). The main effect of time showed a significant difference in diastolic blood pressure over time, $F(1, 44) = 5.444$, $p < 0.0005$, partial $\eta^2 = 0.110$. Post hoc analysis revealed diastolic blood pressure was significantly reduced at week 12 (-3.1 mmHg, $p = 0.02$) compared to baseline (83.96 mmHg, SE 1.6).

3.12. Blood Glucose

There were no significant differences in baseline blood glucose (Table 1) between mRMR (Md = 4.85, *n* = 28) and eRMR participants (Md = 5.30, *n* = 25), U = 392, z = 0.749, *p* = 0.454, r = 0.10. There was no statistically significant interaction between the intervention and time on blood glucose concentration (Table 4), F(1, 43) = 0.112, *p* = 0.739, partial η^2 = 0.003. The main effect of group showed no significant difference in blood glucose concentration between the intervention groups, F(1, 43) = 3.203, *p* = 0.081, partial η^2 = 0.069. The main effect of time showed a statistically significant difference in blood glucose concentration over time, F(1, 43) = 9.120, *p* < 0.0005, partial η^2 = 0.175. Post hoc analysis revealed that blood glucose was significantly lower at week 12 (-0.3 mmol/L, *p* = 0.004) compared to baseline (5.2 mmol/L, SE 0.12) (Figure 3).

3.13. Total Cholesterol

There was a significant difference in baseline total cholesterol between mRMR (Table 1) (Md = 4.74, *n* = 28) and eRMR participants (Md = 3.935, *n* = 22), U = 186, z = -2.385, *p* = 0.017, r = 0.34. There was no significant interaction between the intervention and time on total cholesterol concentration (Table 4), F(1, 41) = 0.584, *p* = 0.449, partial η^2 = 0.014. The main effect of group showed no significant difference in total cholesterol between the intervention groups, F(1, 41) = 1.836, *p* = 0.183, partial η^2 = 0.043. The main effect of time showed no significant difference in total cholesterol over time, F(1, 41) = 0.011, *p* = 0.918, partial η^2 = 0.000 (Figure 4).

3.14. High-Density Lipoprotein

There were no significant differences in baseline HDL (Table 1) between mRMR (Md = 1.34, *n* = 28) and eRMR participants (Md = 1.23, *n* = 23), U = 306, z = -0.303, *p* = 0.762, r = -0.04. There was no statistically significant interaction between the intervention and time on HDL (Table 4), F(1, 42) = 2.196, *p* = 0.146, partial η^2 = 0.050. The main effect of group showed that there was no significant difference in HDL (Figure 5) between the intervention groups F(1, 42) = 0.014, *p* = 0.908, partial η^2 = 0.000. The main effect of time showed a significant difference in HDL concentration over time, F(1, 42) = 4.659, *p* < 0.0005, partial η^2 = 0.100. Post hoc analysis revealed that HDL was significantly lower at week 12 (-0.1 mmol/L, *p* = 0.04) compared to baseline (1.4 mmol/L, SE 0.1).

3.15. Total Cholesterol to High-Density Lipoprotein Ratio

There were no significant differences in baseline TC:HDL (Table 1) between mRMR (Md = 3.4, *n* = 28) and eRMR participants (Md = 3.0, *n* = 21), U = 206, z = -1.771, *p* = 0.077, r = 0.25. There was no statistically significant interaction between the intervention and time on TC:HDL (Table 4), F(1, 40) = 0.028, *p* = 0.869, partial η^2 = 0.001. The main effect of group showed no significant difference in TC:HDL between the intervention groups, F(1, 40) = 3.383, *p* = 0.073, partial η^2 = 0.078. The main effect of time showed no significant difference in TC:HDL over time (Table 2), F(1, 40) = 3.043, *p* = 0.089, partial η^2 = 0.071 (Figure 6).

3.16. Low-Density Lipoprotein

There was a significant difference in baseline LDL (Table 1) between mRMR (Md = 2.77, *n* = 24) and eRMR participants (Md = 2.21, *n* = 19), U = 104, z = -3.033, *p* = 0.002, r = 0.50. There was no statistically significant interaction between the intervention and time on LDL (Table 4), F(1, 34) = 1.194, *p* = 0.282, partial η^2 = 0.034. The main effect of group showed that there was a substantial statistically significant difference LDL between the intervention groups, F(1, 34) = 5.864, *p* = 0.021, partial η^2 = 0.15. Post hoc analysis revealed that LDL was significantly lower in the eRMR group than the mRMR group ($-.491$ mmol/L, *p* = 0.021). The main effect of time showed no statistically significant difference in LDL over time, F(1, 34) = 0.642, *p* = 0.429, partial η^2 = 0.019 (Figure 7).

3.17. Triglycerides

There were no significant differences in baseline triglycerides (Table 1) between mRMR (Md = 1.08, n = 25) and eRMR participants (Md = 1.19, n = 19), U = 278, z = 0.960, p = 0.337, r = 0.14. There was no significant interaction between the intervention and time on triglyceride concentration, $F(1, 36)$ = 0.186, p = 0.669, partial η^2 = 0.005 (Table 4). The main effect of group showed no significant difference in triglyceride concentration between the intervention groups $F(1, 36)$ = 0.242, p = 0.626, partial η^2 = 0.007. The main effect of time showed no significant difference in triglyceride concentration over time, $F(1, 36)$ = 2.028, p = 0.163, partial η^2 = 0.053, (Figure 8).

3.18. Energy Intake

There were no significant differences in baseline energy intake (Table 1) of mRMR (Md = 2195.0, n = 21) and eRMR (Md = 2129.0, n = 19), U = 184, z = -0.420, p = 0.68, r = 0.07. There was no significant interaction between the intervention and time on energy intake (Table 5), $F(1, 28)$ = 0.003, p = 0.956, partial η^2 = 0.000. The main effect of group showed no significant difference in energy intake between the intervention groups, $F(1, 28)$ = 1.225, p = 0.279, partial η^2 = 0.042. The main effect of time showed a substantial statistically significant difference in energy intake at the different time points, $F(1, 28)$ = 14.934, p < 0.0005, partial η^2 = 0.348. Post hoc analysis revealed that energy intake was significantly lower at week 12 compared to baseline (-479 kcal/day, p = 0.001).

3.19. Measured Resting Metabolic Rate

There were no significant differences in baseline measured RMR (Table 1) between mRMR (Md = 1604.00, n = 29) and eRMR participants (Md = 1691.00, n = 23), U = 347, z = 0.249, p = 0.804, r = 0.04. There was no significant interaction between the intervention groups and time on measured RMR (Table 5), $F(2.421, 87.171)$ = 0.284, p = 0.794, partial η^2 = 0.008. The main effect of group showed that there was no significant difference for measured RMR between the intervention groups, $F(1, 36)$ = 0.101, p = 0.752, partial η^2 = 0.003. The main effect of time showed no significant difference for measured RMR over time, $F(2.421, 87.171)$ = 1.525, p = 0.220, partial η^2 = 0.041.

3.20. Respiratory Exchange Ratio

There were no significant differences in baseline measured RER (Table 1) between mRMR (Md = 0.77, n = 29) and eRMR participants (Md = 0.81, n = 23), U = 381, z = 0.876, p = 0.381, r = 0.12. There was no significant interaction between the intervention groups and time on RER (Table 5), $F(2.339, 84.206)$ = 1.112, p = 0.340, partial η^2 = 0.030. The main effect of group showed that there was no significant difference in measured RER between the intervention groups, $F(1, 36)$ = 0.105, p = 0.748, partial η^2 = 0.003. The main effect of time showed no significant difference in RER over time, $F(2.339, 84.206)$ = 1.159, p = 0.324, partial η^2 = 0.031.

3.21. Predicted Resting Metabolic Rate

There were no significant differences in RMR predicted by Mifflin et al. (1990) at baseline (Table 1) between mRMR group (1560.28 $\pm$ 221.71) and eRMR groups (1639.25 $\pm$ 272.21; $t(51)$ = -1.164, p = 0.250, mean difference = -78.97 (95% CI, -215.13 to 57.18). There was no significant interaction between the intervention and time on eRMR (Table 5), $F(1.83, 67.60)$ = 0.215, p = 0.787, partial η^2 = 0.837. The main effect of group showed that there was no statistically significant difference in eRMR between the intervention groups, $F(1, 37)$ = 0.841, p = 0.365, partial η^2 = 0.022. The main effect of time showed a significant difference in eRMR over time, $F(1.83, 84)$ = 12.88, p < 0.0005, partial η^2 = 0.258. Post hoc analysis revealed that eRMR was significantly lower (both < p = 0.0005) at week 6 (1.8%), and week 12 (1.9%) compared to baseline. Post hoc analysis revealed no significant difference between eRMR at week 3 versus baseline (0.3%, p= 1.0). Post hoc analysis revealed that eRMR at week 6 (1.4%) and week 12 (1.6%) was significantly (both < p = 0.023) lower than eRMR at week 3. Post hoc analysis revealed no significant difference between eRMR at week 6 versus week 12 (0.2%, $p = 1.0$).

4. Discussion

The aim of this study was to compare the efficacy of a dietary intervention (mRMR versus eRMR) on weight outcomes in Irish adults aged 50 years and over with overweight and obesity. The primary outcome of this study indicates that employing a reduced-calorie diet using IC to determine energy needs when improving weight outcomes in adults with overweight and obesity is equal to employing a reduced-calorie diet based on the Mifflin et al. [28] prediction equation. Following the study period, a significant ($p < 0.05$) reduction in body weight was observed in both mRMR (-4.2% of initial body weight) and eRMR (-3.2%) groups. However, there were no significant ($p \geq 0.05$) differences between groups. Overall, 20.8% and 17.4% of mRMR and eRMR participants, respectively, experienced clinically meaningful weight reduction. One participant in the eRMR group experienced a 10–14.9% weight reduction, and one participant in the mRMR group experienced a more than 15% weight reduction. Rapid weight loss may be a sign of underlying health conditions or chronic disease. No health condition, disease or illness was identified prior to or during the intervention that may be attributed to rapid or unintentional weight loss. From the one-to-one consultations, it can be assumed that the observed weight loss may be attributed to successful adherence to diet and lifestyle modifications. While a secondary analysis of data assessing biological sex differences in weight variation within the mRMR and eRMR groups was not investigated in the present study, a previous study [46] with similar participants investigated gender differences in weight and BMI variation in response to a dietary intervention based on measured RMR using IC and equations. No statistically significant differences in body weight and BMI variation between the two IC and no IC groups were found between males and females (three-way interaction time by treatment by gender: $p = 0.16$ for BMI, $p = 0.11$ for weight), although a trend to a greater weight loss in females was observed in both groups. Secondary outcome measures revealed a significant reduction ($p \leq 0.05$) in BMI, WC and muscle mass in both groups. Differences observed between groups were not significant ($p \geq 0.05$). There were no significant ($p \geq 0.05$) differences in percent body fat over time or between groups. Both groups experienced a significant ($p \leq 0.05$) reduction in systolic and diastolic BP following the intervention period. Blood glucose and triglycerides were significantly ($p \leq 0.05$) lower in both groups and there was no significant ($p \geq 0.05$) difference observed between groups. There was no significant ($p \geq 0.05$) difference for total cholesterol or TC:HDL over time or between groups. HDL concentration was significantly ($p < 0.05$) lower in both groups with no significant difference ($p \geq 0.05$) between groups. No significant difference ($p \geq 0.05$) was observed for LDL over the study period. However, there were significant ($p < 0.05$) differences in LDL between groups at baseline, and 47% of participants in the eRMR group displayed an upward trend in LDL over the study period (Figure 3).

In contrast, previous studies reported significant between-group differences when comparing similar dietary interventions (i.e., prediction equations versus metabolic based) in a comparable population [46,47]. Participants following a nutrition plan based on eRMR for a period of 90 days experienced a 2% weight reduction compared to -4.5% when following a diet based on mRMR [47]. The between-group differences may be explained by the methical differences used to estimate RMR. Massarini et al. [46] employed the Harris and Benedict [48] prediction equation while McDoniel et al. [47] employed the American College of Chest Physicians (ACCP) prediction equation ($25 \times$ baseline body weight (kg) $- 250$ to 500 kcal/day) [5,49]. The application of prediction equations provides a source of variability as often they are utilised for a population which they were not originally developed for, thus resulting in reduced accuracy among specific populations [50,51]. Later work conducted by McDoniel et al. [52] in a similar population reported similar results to the present study. McDoniel et al. [52] conducted a 24 week randomised controlled trial, where usual care participants received a fixed low-calorie diet (i.e., 1200 kcal/day for females and 1600 kcal/day for males, respectively) and participants in the metabolic diet (MD) group received an individualised nutrition plan based on mRMR. McDoniel and Hammond [52] reported a significant reduction in body weight at week 12, but similar to the current study observed no significant differences between the intervention groups. When comparing usual care practice to eRMR, participants following the fixed-calorie diet experienced a greater weight reduction (1.3%). Participants in the usual care group experienced a 4.5% reduction in bodyweight compared to the eRMR group (−3.2%) in the present study. It is reasonable to assume that the greater weight reduction observed by McDoniel and Hammond [52] in the usual care group may be attributed to a greater energy deficit compared to participants in the eRMR group. Usual care participants were prescribed approximately 300–500 kcal/day less than eRMR participants (Table 7) (females: 1500 ± 141 kcal/day; males: 2100 ± 236 kcal/day). Standardised hypocaloric balanced diets are designed to facilitate a 0.5–1.0 kg per week weight reduction by consuming approximately 500 kcal/day less than required for weight maintenance. This is based on the assumption that a negative energy balance of 7700 kcal is required for a 1 kg reduction in body weight. Therefore, an energy deficit of 3500 kcal/week should result in a 0.5 kg/week weight reduction. McDoniel et al. [53] prescribed similar energy intakes to that described in this study and observed comparable results. Participants in the self-monitoring and RMR technology (SMART) group received a nutrition plan based on mRMR and usual care participants were prescribed a standardised ad libitum diet (females:1200 kcal/day; males:1600 kcal/day). Participants in the SMART group experienced a 3.9% weight reduction, slightly lower than but similar to participants in the present study (−4.2%) employing comparable metabolic based diets (mRMR). Energy prescription was based on RMR measured by IC, which may explain the similarities. For instance, women in the mRMR and SMART group were prescribed 1584 ± 271 kcal/day and 1656 ± 334 kcal/day, respectively, while men in the mRMR and SMART group were prescribed 2180 ± 253 kcal/day and 2296 ± 565 kcal/day, respectively. The current study observed a weight gain equal to or less than 2% of initial body weight in 20.8% and 30.4% of participants in the mRMR and eRMR groups, respectively. Factors possibly contributing to an increase in body weight despite a reduced energy prescription include difficulty in adopting positive behaviour change to support dietary changes and thus influence weight.

Secondary outcomes of this study support previous research demonstrating that modest weight reduction lowers blood pressure, triglycerides and glucose [54]. A reduction in systolic (mRMR −2.8 ± 1.1 mmHg; eRMR: −7.0 ± 4.3 mmHg) and diastolic (mRMR: −3.3 ± 2.6; eRMR: 2.9 ± 0.0 mmHg) blood pressure, triglycerides and glucose was observed (Tables 2 and 4). These outcomes may be attributed to components of the dietary intervention which emphasise high fruit and vegetable consumption and intake of wholegrains, while reduced sodium and saturated fat intake, and limited intake of energy dense foods. A high intake of vegetables and fruit is associated with reduced blood pressure and a lower risk of CVD. Furthermore, dietary fibre intake and consumption of wholegrain products are linked to a lower risk of diabetes and reduced diastolic blood pressure, while lowering sodium intake reduces blood pressure [55]. Similar to Zinn et al. [56] a non-significant upward and downward trend in LDL was observed in the eRMR and mRMR groups, respectively, with 47% of participants in the eRMR group displaying an upward trend of LDL and 31.6% of participants in the mRMR group showing a downward trend in LDL (Figure 8). A systematic literature review evaluating the effect of energy restriction diets on weight loss outcomes in adults with overweight and obesity reported significant reductions in FFM or lean body mass in six of the included studies (n = 216) [57]. Muscle mass was significantly reduced at week 3 (0.8 kg) and week 12 (2.1 kg) compared to baseline (Table 2). Loss of FFM is unfavourable for numerous reasons including the impact on metabolic health, functional capacity, i.e., the ability to carry out activities of daily living and the increased risk of injury associated with reduced functional capacity. Greater FFM is linked to a higher metabolic rate, which is advantageous for weight reduction. In an effort to offset the potential loss of FFM, the present study encouraged adherence to current national physical activity guidelines. National physical activity guidelines for older adults recommend at least 30 min a day of moderate-intensity activity on five days a week, or 150 min per week. The addition of moderate-intensity aerobic exercise, primarily walking, to intentional weight loss has been shown to attenuate the loss of muscle mass in older adults with overweight and obesity [58]. Aerobic activity, muscle-strengthening and balance activities form part of the recommendations for adults (Get Ireland Active, HSE). Despite the effort made by participants in this study to attenuate the effects of lean mass loss, a significant decrease was observed.

There was no significant difference in RMR measured by IC at baseline or across the intervention period for both groups (Tables 1 and 5). This observation supports previous studies that suggest resting energy expenditure remains relatively stable following modest energy deficit diet. Severe energy restriction is associated with greater loss in metabolically active tissue, and thus results in metabolic adaptation [59]. A recent systematic review demonstrated that resistance exercise was effective for increasing RMR. However, the same was not apparent for aerobic exercise [60]. This is an important finding given that the addition of physical activity to weight management programmes is an accepted practice to encourage energy expenditure. It is unclear whether the addition of resistance exercise specifically to weight reduction programmes may attenuate metabolic adaptation by increasing RMR. Future research is required to determine the extent to which resistance exercise affects metabolic adaptation of RMR related to energy restriction.

RMR estimated using the Mifflin et al. [28] prediction equation was lower than actual RMR for both groups (Table 5). eRMR was significantly lower at weeks 6 and 12 compared to baseline for participants in both the mRMR and eRMR groups. If eRMR was lower than actual mRMR, then participants following a diet based on estimated energy needs would receive a dietary plan underestimating calorie requirement and, therefore, greater weight loss would be anticipated. When separated out, male and female analysis of energy intake estimated using 3-day food diaries indicate that both male and female eRMR participants were under their prescribed intake at week 12, leading to further deficit than required (Table 7). Given that similar weight loss was evident between the eRMR and mRMR groups, this would suggest that the same weight loss may occur with a higher-calorie plan.

Furthermore, results from this study suggest energy intake closely matched to mRMR results in higher level of dietary adherence in male participants. Estimated male energy intake (2052 ± 597 kcal) from 3-day food diaries was very similar to energy intake prescribed using mRMR which is advantageous in weight reduction strategies. Strong adherence level is associated with successful outcomes in weight management [61]. A recent study investigating sociocultural gender factors which influence food behaviours, such as dietary preferences and adherence, reported that men were more adherent to a healthy low-carbohydrate (HLC) than women ($p = 0.02$) vs. healthy low-fat (HLF) diet [62]. Another possible reason for this may be the addition of regular one-to-one support provided by the researcher, where goal setting and self-monitoring of food intake were encouraged using motivational interviewing techniques and behavioural change skills. This may not be the case for women as estimated energy intake for females (1736 ± 473 kcal) was higher than prescribed and indicates that additional support may be warranted to aid adherence to weight management programmes for females. Assessment of energy intake is often unreliable, particularly in individuals classified as overweight or affected by obesity [14]. Therefore, a greater frequency of dietary self-monitoring, which is associated with greater weight loss success, may benefit females [63].

4.1. Strengths of Study

This research was unique in that it was the first study to compare RMR estimated by the Mifflin et al. (1990) prediction equation using actual body weight to inform dietary prescription directly to dietary prescription based on measured RMR value in this specific population group, namely Irish adults aged 50 years and over with a BMI ≥ 25 kg/m^2 and of Caucasian ethnicity. Thus, the results of this study may be extrapolated to Irish adults aged 50 years or over with a BMI greater than 25 kg/m^2. A major strength of this research was the level of adherence to the intervention, as a 93% retention rate was observed. For power based on weight reduction calculations and allowing for a 20% drop out, 46 people were required in total (approximately 23 in each intervention group). Fifty-two participants completed this study, thus allowing for statistical significance. Extensive support was provided by the researcher to all participants during this study, with regular one-to-one support, individualised nutrition coaching and accessible to participants should they have any questions or concerns. The researcher was certified in motivational interviewing and behaviour change skills, which, alongside a non-judgemental patient-centred approach, enhanced the consultation process and gave the participants an opportunity to engage in the conversation about health concerns relating to their weight. McGowan [64] highlighted the importance of strong communication skills, avoiding stigmatisation and the appropriate use of person-first language as imperative to successfully engaging patients. This level of engagement allowed for individual adjustments to be made where necessary and highlights that, irrespective of how the calorie deficit is achieved, it is the individualised and tailored approach that is the most important factor for retention and thus achieving a successful outcome. Further strengths of this study include consistent data collection by the same researcher, which reduces potential measurement error.

4.2. Limitations

There are a few limitations to this study. A secondary analysis of data for biological sex differences in weight variation within the mRMR and eRMR groups, or other sex-related factors, such as genotype, hormones, metabolic syndrome, or psychosocial factors that may affect either adherence to dietary intervention or weight reduction response were not investigated. Aronica et al. [62] highlighted the need for such analysis while acknowledging that a limited amount of weight reduction studies demonstrated sufficient power to compare the effects of energy restriction diets on weight outcomes in women vs. men. The present study duration was 12 weeks, thus it is unknown whether participants in the mRMR or eRMR groups maintained weight reduction over a longer period. While qualitative measures were collected at baseline to estimate the energy expenditure associated with physical activity and to inform the subsequent dietary intervention accurately, measuring energy expenditure associated with physical activity throughout this study using measurement devices such as the gold-standard doubly labelled water method or wearable monitors such as accelerometers and movement sensors was beyond the scope of this study. Some participants may have become more physically activity as this study progressed and so the contributory mechanism of physical activity energy expenditure to weight outcome remains speculative. Finally, all participants in this study were of a specific age range ($\geq$50 years), BMI class (25 kg/m^2) and ethnic group (Caucasian) and thus caution should be used when extrapolating these results to other population groups.

4.3. Future Direction

Based on these data, the use of RMR technology demonstrates promise for effective weight reduction outcomes. Future research is needed to better understand the efficacy of RER and substrate utilisation information in tailoring dietary prescription. To address the gap in the literature as identified by Aronica et al. [62] biological sex differences such as body composition and metabolism should be investigated among research participants to compare the effects of diets with energy prescription based on IC or prediction equations on weight outcomes. Furthermore, sociocultural gender factors which influence food behaviours such as dietary preferences and adherence should be explored. The high retention rate (93%) demonstrated in this study would suggest that participants are more likely to adhere to a modest calorie-deficit nutrition programme with regular support. The effects over longer time frames are less clear and future research is required to investigate whether compliance to a modest calorie deficit is more suited to older adult population long term. Future research should accurately measure physical activity energy expenditure to determine its contribution to weight outcomes. Additional research is needed to determine the extent to which resistance exercise affects metabolic adaptation of RMR related to energy restriction.

5. Conclusions

In conclusion, the results of this study suggest that a reduced-energy diet based on mRMR or eRMR facilitates weight reduction in adults aged $\geq$ 50 years over the short term (12 weeks) and favours a reduction in blood pressure, triglycerides and glucose, thus reducing CVD risk factors. Overall, 20.8% mRMR and 17.4% of eRMR participants experienced clinically meaningful (i.e., $\geq$5% of initial weight) weight reduction. Moreover, dietary approaches that entail modest calorie deficit combined with individual counselling using MI and behaviour change skills that support and encourage small behaviour changes may be effective in short-term (up to 12 weeks) adult (aged $\geq$ 50 years) weight management.

Supplementary Materials: The following are available online at https://www.mdpi.com/article/10.3390/nu13041229/s1. Supplementary Figure S1 Informed consent was obtained from all subjects involved in the study

Author Contributions: Conceptualisation, C.C. and L.R.; methodology, C.C., L.R. and M.M. software, C.C.; validation, C.C. and L.R.; formal analysis, C.C.; investigation, C.C.; resources, L.R. and C.C.; data curation, C.C.; writing—original draft preparation, C.C.; writing—review and editing, C.C., E.D., M.M. and L.R.; visualisation, C.C.; supervision, L.R.; project administration, C.C. and L.R.; funding acquisition, L.R. All authors have read and agreed to the published version of the manuscript.

Funding: C.C. was the recipient of a scholarship awarded by the Irish Smart Ageing Exchange (ISAX) and the Research and Innovation Strategic Endowment (RISE) scholarship scheme (GMIT).

Institutional Review Board Statement: The study was conducted according to the guidelines of the Declaration of Helsinki, and approved by the Research Ethics Committee of Galway Mayo Institute of Technology (GMIT), Ireland (protocol code RSC_AC230119 and date of approval 23 January 2019). Informed consent was obtained from all subjects involved in the study.

Informed Consent Statement: All participants provided written informed consent prior to their inclusion in this study.

Data Availability Statement: The study was conducted in accordance with the Data Protection ACT, 2018 and approved by the Institute's Data Protection Officer. Researchers seeking the analysis dataset for this work should submit requests to the corresponding author.

Acknowledgments: The authors are grateful to all the participants for their time in contributing to this research. We acknowledge and thank the Department of Natural Sciences, School of Science and Computing at GMIT for the use of laboratory equipment and facilities. The authors would like to acknowledge and thank Nóra Ní Fhlannagáin RD and Emma Finnegan for their contribution to the dietary intervention.

Conflicts of Interest: The authors declare no conflict of interest.

References

1. Donoghue, O.; Dooley, C.; Feeney, J.; Finucane, C.; Hudson, E.; Kenny, R.A.; Leahy, S.; McCrory, C.; McGarrigle, C.; McHugh, S.; et al. *The Over 50s in a Changing Ireland Economic Circumstances, Health and Well-Being*; The Irish Longitudinal Study on Ageing: Dublin, Ireland, 2014; pp. 86–103.
2. Cowley, M.A.; Brown, W.A.; Considine, R.V. Obesity: The problem and its management. In *Endocrinology: Adult & Pediatric*; Elsevier: Amsterdam, The Netherlands, 2016; pp. 468–478.
3. Soysal, P.; Bulut, E.A.; Yavuz, I.; Isik, A.T. Decreased Basal Metabolic Rate Can Be an Objective Marker for Sarcopenia and Frailty in Older Males. *J. Am. Med Dir. Assoc.* **2019**, *20*, 58–63. [CrossRef] [PubMed]
4. Lee, T.C.; Jin, Z.; Homma, S.; Nakanishi, K.; Elkind, M.S.; Rundek, T.; Tugcu, A.; Matsumoto, K.; Sacco, R.L.; Di Tullio, M.R. Changes in Left Ventricular Mass and Geometry in the Older Adults: Role of Body Mass and Central Obesity. *J. Am. Soc. Echocardiogr.* **2019**, *32*, 1318–1325. [CrossRef]
5. Tsigos, C.; Hainer, V.; Basdevant, A.; Finer, N.; Fried, M.; Mathus-Vliegen, E.; Micic, A.; Maislos, M.; Roman, G.; Schutz, Y.; et al. Management of obesity in adults: European clinical practice guidelines. *Obes. Facts* **2008**, *1*, 106–116. [CrossRef] [PubMed]

6. Stegenga, H.; Haines, A.; Jones, K.; Wilding, J.; On behalf of the Guideline Development Group. Identification, assessment, and management of overweight and obesity: Summary of updated NICE guidance. *BMJ* **2014**, *349*, g6608. [CrossRef] [PubMed]
7. Yumuk, V.; Tsigos, C.; Fried, M.P.; Schindler, K.; Busetto, L.; Micic, D.; Toplak, H. European Guidelines for Obesity Management in Adults. *Obes. Facts* **2015**, *8*, 402–424. [CrossRef]
8. Villareal, D.T.; Apovian, C.M.; Kushner, R.F.; Klein, S. Obesity in Older Adults: Technical Review and Position Statement of the American Society for Nutrition and NAASO, The Obesity Society. *Obes. Res.* **2005**, *13*, 1849–1863. [CrossRef]
9. National Institute for Health Care Excellence. Overview | Obesity: Identification, Assessment and Management. Available online: https://www.nice.org.uk/guidance/cg189 (accessed on 7 April 2021).
10. Raynor, H.A.; Champagne, C.M. Position of the Academy of Nutrition and Dietetics: Interventions for the Treatment of Overweight and Obesity in Adults. *J. Acad. Nutr. Diet.* **2016**, *116*, 129–147. [CrossRef]
11. Von Loeffelholz, C.; Birkenfeld, A. The Role of Non-Exercise Activity Thermogenesis in Human Obesity, in Endotext. 2018. Available online: https://www.ncbi.nlm.nih.gov/books/NBK279077/ (accessed on 7 April 2021).
12. Levine, J.A. Measurement of energy expenditure. *Public Heal. Nutr.* **2005**, *8*, 1123–1132. [CrossRef] [PubMed]
13. Volp, A.C.P.; De Oliveira, F.C.E.; Alves, R.D.M.; Esteves, E.A.; Bressan, J. Energy expenditure: Components and evaluation methods. *Nutr. Hosp.* **2011**, *26*, 430–440.
14. Gupta, R.D.; Ramachandran, R.; Padmanaban Venkatesan, S.A.; Joseph, M.; Thomas, N. Indirect calorimetry: From bench to bedside. *Indian J. Endocrinol. Metab.* **2017**, *21*, 594.
15. Miles-Chan, J.L.; Dulloo, A.G.; Schutz, Y. Fasting substrate oxidation at rest assessed by indirect calorimetry: Is prior die-tary macronutrient level and composition a confounder? *Int. J. Obes.* **2015**, *39*, 1114–1117. [CrossRef]
16. Purdom, T.; Kravitz, L.; Dokladny, K.; Mermier, C. Understanding the factors that effect maximal fat oxidation. *J. Int. Soc. Sports Nutr.* **2018**, *15*, 1–10. [CrossRef]
17. Kelley, D.E. Skeletal muscle fat oxidation: Timing and flexibility are everything. *J. Clin. Investig.* **2005**, *115*, 1699–1702. [CrossRef]
18. Houmard, J.A. Intramuscular lipid oxidation and obesity. *Am. J. Physiol. Integr. Comp. Physiol.* **2008**, *294*, R1111–R1116. [CrossRef] [PubMed]
19. Lanzi, S.; Codecasa, F.; Cornacchia, M.; Maestrini, S.; Salvadori, A.; Brunani, A.; Malatesta, D. Fat Oxidation, Hormonal and Plasma Metabolite Kinetics during a Submaximal Incremental Test in Lean and Obese Adults. *PLoS ONE* **2014**, *9*, e88707. [CrossRef] [PubMed]
20. Rezuş, E.; Burlui, A.; Cardoneanu, A.; Rezuş, C.; Codreanu, C.; Pârvu, M.; Zota, G.R.; Tamba, B.I. Inactivity and Skeletal Muscle Metabolism: A Vicious Cycle in Old Age. *Int. J. Mol. Sci.* **2020**, *21*, 592. [CrossRef] [PubMed]
21. Marra, M.; Scalfi, L.; Contaldo, F.; Pasanisi, F. Fasting Respiratory Quotient as a Predictor of Long-Term Weight Changes in Non-Obese Women. *Ann. Nutr. Metab.* **2004**, *48*, 189–192. [CrossRef] [PubMed]
22. Fabbri, E.; Shaffer, N.C.; Gonzalez-Freire, M.; Shardell, M.D.; Zoli, M.; Studenski, S.A.; Ferrucci, L. Early body composition, but not body mass, is associated with future accelerated decline in muscle quality. *J. Cachex Sarcopenia Muscle* **2017**, *8*, 490–499. [CrossRef]
23. Ruiz, J.R.; Ortega, F.B.; Rodríguez, G.; Alkorta, P.; Labayen, I. Validity of Resting Energy Expenditure Predictive Equations before and after an Energy-Restricted Diet Intervention in Obese Women. *PLoS ONE* **2011**, *6*, e23759. [CrossRef]
24. Alves, V.G.F.; da Rocha EE, M.; Gonzalez, M.C.; da Fonseca RB, V.; do Nascimento Silva, M.H.; Chiesa, C.A. Assessement of resting energy expenditure of obese patients: Comparison of indirect calorimetry with formulae. *Clin. Nutr.* **2009**, *28*, 299–304. [CrossRef]
25. Madden, A.; Mulrooney, H.M.; Shah, S. Estimation of energy expenditure using prediction equations in overweight and obese adults: A systematic review. *J. Hum. Nutr. Diet.* **2016**, *29*, 458–476. [CrossRef]
26. Tallon, J.M.; Narciso, J.; Saavedra, R.; Silva, A.J.; Barros, A.; da Costa, A.M. Comparison of predictive equations for resting metabolic rate in Portuguese women. *Motricidade* **2020**, *16*, 74–83.
27. Weijs, P.J.M. Validity of predictive equations for resting energy expenditure in US and Dutch overweight and obese class I and II adults aged 18–65 y. *Am. J. Clin. Nutr.* **2008**, *88*, 959–970. [CrossRef]
28. Mifflin, M.D.; Jeor, S.T.S.; Hill, L.A.; Scott, B.J.; Daugherty, S.A.; Koh, Y.O. A new predictive equation for resting energy expenditure in healthy individuals. *Am. J. Clin. Nutr.* **1990**, *51*, 241–247. [CrossRef] [PubMed]
29. Frankenfield, D.; Roth-Yousey, L.; Compher, C. Comparison of Predictive Equations for Resting Metabolic Rate in Healthy Nonobese and Obese Adults: A Systematic Review. *J. Am. Diet. Assoc.* **2005**, *105*, 775–789. [CrossRef] [PubMed]
30. Jeor, S.T.S.; Cutter, G.R.; Perumean-Chaney, S.E.; Hall, S.J.; Herzog, H.; Bovee, V. The Practical Use of Charts to Estimate Resting Energy Expenditure in Adults. *Top. Clin. Nutr.* **2004**, *19*, 51–56. [CrossRef]
31. Zampino, M.; Semba, R.D.; Adelnia, F.; Spencer, R.G.; Fishbein, K.W.; Schrack, J.A.; Simonsick, E.M.; Ferrucci, L. Greater Skeletal Muscle Oxidative Capacity Is Associated With Higher Resting Metabolic Rate: Results From the Baltimore Longitudinal Study of Aging. *J. Gerontol. Ser. A Boil. Sci. Med Sci.* **2020**, *75*, 2262–2268. [CrossRef] [PubMed]
32. Knuth, N.D.; Johannsen, D.L.; Tamboli, R.A.; Marks-Shulman, P.A.; Huizenga, R.; Chen, K.Y.; Abumrad, N.N.; Ravussin, E.; Hall, K.D. Metabolic adaptation following massive weight loss is related to the degree of energy imbalance and changes in circulating leptin. *Obesity* **2014**, *22*, 2563–2569. [CrossRef] [PubMed]

33. Geisler, C.; Braun, W.; Pourhassan, M.; Schweitzer, L.; Glüer, C.-C.; Bosy-Westphal, A.; Müller, M.J. Age-Dependent Changes in Resting Energy Expenditure (REE): Insights from Detailed Body Composition Analysis in Normal and Overweight Healthy Caucasians. *Nutrients* **2016**, *8*, 322. [CrossRef]

34. Abizanda, P.; Romero, L.; Sánchez-Jurado, P.M.; Ruano, T.F.; Ríos, S.S.; Sánchez, M.F. Energetics of Aging and Frailty: The FRADEA Study. *J. Gerontol. Ser. A Boil. Sci. Med Sci.* **2016**, *71*, 787–796. [CrossRef]

35. Redman, L.M.; Heilbronn, L.K.; Martin, C.K.; De Jonge, L.; Williamson, D.A.; Delany, J.P.; Ravussin, E.; Pennington CALERIE Team. Metabolic and Behavioral Compensations in Response to Caloric Restriction: Implications for the Maintenance of Weight Loss. *PLoS ONE* **2009**, *4*, e4377. [CrossRef]

36. Galgani, J.E.; Santos, J.L. Insights about weight loss-induced metabolic adaptation. *Obesity* **2016**, *24*, 277–278. [CrossRef]

37. Redman, L.M.; Smith, S.R.; Burton, J.H.; Martin, C.K.; Il'Yasova, D.; Ravussin, E. Metabolic Slowing and Reduced Oxidative Damage with Sustained Caloric Restriction Support the Rate of Living and Oxidative Damage Theories of Aging. *Cell Metab.* **2018**, *27*, 805–815.e4. [CrossRef]

38. Most, J.; Redman, L.M. Impact of calorie restriction on energy metabolism in humans. *Exp. Gerontol.* **2020**, *133*, 110875. [CrossRef]

39. Stiegler, P.; Cunliffe, A. The Role of Diet and Exercise for the Maintenance of Fat-Free Mass and Resting Metabolic Rate During Weight Loss. *Sports Med.* **2006**, *36*, 239–262. [CrossRef] [PubMed]

40. Johnstone, A.M.; Murison, S.D.; Duncan, J.S.; Rance, K.A.; Speakman, J.R. Factors influencing variation in basal metabolic rate include fat-free mass, fat mass, age, and circulating thyroxine but not sex, circulating leptin, or triiodothyronine. *Am. J. Clin. Nutr.* **2005**, *82*, 941–948. [CrossRef] [PubMed]

41. Craig, C.L.; Marshall, A.L.; Sjöström, M.; Bauman, A.E.; Booth, M.L.; Ainsworth, B.E.; Pratt, M.; Ekelund, U.; Yngve, A.; Sallis, J.F.; et al. International Physical Activity Questionnaire: 12-Country Reliability and Validity. *Med. Sci. Sports Exerc.* **2003**, *35*, 1381–1395. [CrossRef]

42. Nutritics. References. 2020. Available online: https://www.nutritics.com/p/home (accessed on 7 April 2021).

43. Food Safety Authority of Ireland. *Healthy Eating, Food Safety and Food Legislation—A Guide Supporting the Healthy Ireland Food Pyramid*; Food Safety Authority of Ireland Healthy Ireland, Department of Health, the Health Service Executive (HSE), Eds.; The Food Safety Authority of Ireland (FSAI): Dublin, Ireland, 2019.

44. Healthy Ireland and Health Service Executive. *Get Ireland Walking: Walk for Health | Walk for Fun | Walk for Life*; Healthy Ireland, Health Service Executive: Dublin, Ireland, 2019.

45. Eston, R.; Reilly, T. *Kinanthropometry and Exercise Physiology Laboratory Manual: Test, Procedures and Data*, 3rd ed.; Volume One: Anthropometry; Routledge: Oxon, UK, 2009.

46. Massarini, S.; Ferrulli, A.; Ambrogi, F.; Macrì, C.; Terruzzi, I.; Benedini, S.; Luzi, L. Routine resting energy expenditure measurement increases effectiveness of dietary intervention in obesity. *Acta Diabetol.* **2018**, *55*, 75–85. [CrossRef] [PubMed]

47. McDoniel, S.O.; Nelson, H.A.; Thomson, C.A. Employing RMR Technology in a 90-Day Weight Control Program. *Obes. Facts* **2008**, *1*, 298–304. [CrossRef] [PubMed]

48. Harris, J.A.; Benedict, F.G. A Biometric Study of Human Basal Metabolism. *Proc. Natl. Acad. Sci. USA* **1918**, *4*, 370–373. [CrossRef]

49. Cerra, F.B.; Benitez, M.R.; Blackburn, G.L.; Irwin, R.S.; Jeejeebhoy, K.; Katz, D.P.; Pingleton, S.K.; Pomposelli, J.; Rombeau, J.L.; Shronts, E.; et al. Applied Nutrition in ICU Patients: A Consensus Statement of the American College of Chest Physicians. *Chest* **1997**, *111*, 769–778. [CrossRef]

50. Sabounchi, N.S.; Rahmandad, H.; Ammerman, A. Best-fitting prediction equations for basal metabolic rate: Informing obesity interventions in diverse populations. *Int. J. Obes.* **2013**, *37*, 1364–1370. [CrossRef]

51. Flack, K.D.; Siders, W.A.; Johnson, L.; Roemmich, J.N. Cross-Validation of Resting Metabolic Rate Prediction Equations. *J. Acad. Nutr. Diet.* **2016**, *116*, 1413–1422. [CrossRef]

52. McDoniel, S.O.; Hammond, R.S. A 24-week randomised controlled trial comparing usual care and metabolic-based diet plans in obese adults. *Int. J. Clin. Pr.* **2010**, *64*, 1503–1511. [CrossRef]

53. McDoniel, S.O.; Wolskee, P.; Shen, J. Treating obesity with a novel hand-held device, computer software program, and In-ternet technology in primary care: The SMART motivational trial. *Patient Educ. Couns.* **2010**, *79*, 185–191. [CrossRef] [PubMed]

54. Harsha, D.W.; Bray, G.A. Response to Weight Loss and Blood Pressure Control: The Pro Side. *Hypertension* **2008**, *51*, 1420–1425. [CrossRef]

55. Kromhout, D.; Spaaij, C.J.K.; De Goede, J.; Weggemans, R.M. The 2015 Dutch food-based dietary guidelines. *Eur. J. Clin. Nutr.* **2016**, *70*, 869–878. [CrossRef] [PubMed]

56. Zinn, C.; McPhee, J.; Harris, N.K.; Williden, M.; Prendergast, K.; Schofield, G. A 12-week low-carbohydrate, high-fat diet improves metabolic health outcomes over a control diet in a randomised controlled trial with overweight defence force personnel. *Appl. Physiol. Nutr. Metab.* **2017**, *42*, 1158–1164. [CrossRef]

57. Davis, C.S.; Clarke, R.; Coulter, S.N.; Rounsefell, K.N.; Walker, R.E.; Rauch, C.; Huggins, C.E.; Ryan, L. Intermittent energy restriction and weight loss: A systematic review. *Eur. J. Clin. Nutr.* **2015**, *70*, 292–299. [CrossRef] [PubMed]

58. Chomentowski, P.; Dubé, J.J.; Amati, F.; Stefanovic-Racic, M.; Zhu, S.; Toledo, F.G.; Goodpaster, B.H. Moderate exercise attenuates the loss of skeletal muscle mass that occurs with intentional caloric re-striction–induced weight loss in older, overweight to obese adults. *J. Gerontol. Ser. A Biomed. Sci. Med Sci.* **2009**, *64*, 575–580. [CrossRef]

59. Trexler, E.T.; Smith-Ryan, A.E.; Norton, L.E. Metabolic adaptation to weight loss: Implications for the athlete. *J. Int. Soc. Sports Nutr.* **2014**, *11*, 7. [CrossRef] [PubMed]

60. MacKenzie-Shalders, K.; Kelly, J.T.; So, D.; Coffey, V.G.; Byrne, N.M. The effect of exercise interventions on resting metabolic rate: A systematic review and me-ta-analysis. *J. Sports Sci.* **2020**, *38*, 1635–1649. [CrossRef] [PubMed]
61. Gibson, A.A.; Sainsbury, A. Strategies to Improve Adherence to Dietary Weight Loss Interventions in Research and Re-al-World Settings. *Behav. Sci.* **2017**, *7*, 44. [CrossRef] [PubMed]
62. Aronica, L.; Rigdon, J.; Offringa, L.C.; Stefanick, M.L.; Gardner, C.D. Examining differences between overweight women and men in 12-month weight loss study comparing healthy low-carbohydrate vs. low-fat diets. *Int. J. Obes.* **2021**, *45*, 225–234. [CrossRef]
63. Harvey, J.; Krukowski, R.; Priest, J.; West, D. Log Often, Lose More: Electronic Dietary Self-Monitoring for Weight Loss. *Obesity* **2019**, *27*, 380–384. [CrossRef] [PubMed]
64. McGowan, B.M. A Practical Guide to Engaging Individuals with Obesity. *Obes. Facts* **2016**, *9*, 182–192. [CrossRef]

Article

Iron Deficiency without Anemia Decreases Physical Endurance and Mitochondrial Complex I Activity of Oxidative Skeletal Muscle in the Mouse

Emmanuel Rineau [1,2,*], Naïg Gueguen [1,3], Vincent Procaccio [1,3], Franck Geneviève [4], Pascal Reynier [1,3], Daniel Henrion [1] and Sigismond Lasocki [1,2]

[1] MitoVasc Institut, UMR CNRS 6015—INSERM 1083, University of Angers, 49100 Angers, France; NaGueguen@chu-angers.fr (N.G.); vincent.procaccio@univ-angers.fr (V.P.); pascal.reynier@univ-angers.fr (P.R.); daniel.henrion@univ-angers.fr (D.H.); silasocki@chu-angers.fr (S.L.)

[2] Department of Anesthesia and Critical Care, University Hospital of Angers, 49100 Angers, France

[3] Department of Biochemistry and Genetics, University Hospital of Angers, 49100 Angers, France

[4] Department of Hematology, University Hospital of Angers, 49100 Angers, France; frgenevieve@chu-angers.fr

* Correspondence: Emmanuel.Rineau@chu-angers.fr; Tel.: +33-2-41-35-36-35

Citation: Rineau, E.; Gueguen, N.; Procaccio, V.; Geneviève, F.; Reynier, P.; Henrion, D.; Lasocki, S. Iron Deficiency without Anemia Decreases Physical Endurance and Mitochondrial Complex I Activity of Oxidative Skeletal Muscle in the Mouse. *Nutrients* **2021**, *13*, 1056. https://doi.org/10.3390/nu13041056

Academic Editor: Arie Nieuwenhuizen

Received: 27 February 2021
Accepted: 19 March 2021
Published: 24 March 2021

Publisher's Note: MDPI stays neutral with regard to jurisdictional claims in published maps and institutional affiliations.

Abstract: Iron deficiency (ID), with or without anemia, is responsible for physical fatigue. This effect may be linked to an alteration of mitochondrial metabolism. Our aim was to assess the impact of ID on skeletal striated muscle mitochondrial metabolism. Iron-deficient non-anemic mice, obtained using a bloodletting followed by a low-iron diet for three weeks, were compared to control mice. Endurance was assessed using a one-hour submaximal exercise on a Rotarod device and activities of mitochondrial complexes I and IV were measured by spectrophotometry on two types of skeletal striated muscles, the soleus and the quadriceps. As expected, ID mice displayed hematologic markers of ID and reduced iron stores, although none of them were anemic. In ID mice, endurance was significantly reduced and activity of the respiratory chain complex I, normalized to citrate synthase activity, was significantly reduced in the soleus muscle but not in the quadriceps. Complex IV activities were not significantly different, neither in the soleus nor in the quadriceps. We conclude that ID without anemia is responsible for impaired mitochondrial complex I activity in skeletal muscles with predominant oxidative metabolism. These results bring pathophysiological support to explain the improved physical activity observed when correcting ID in human. Further studies are needed to explore the mechanisms underlying this decrease in complex I activity and to assess the role of iron therapy on muscle mitochondrial metabolism.

Keywords: iron deficiency; striated skeletal muscle; physical capacity; fatigue; mitochondrial metabolism; complex I

1. Introduction

Fatigue is one of the main symptoms of iron deficiency (ID), even in absence of anemia. Iron deficiency-related fatigue may present as a mental or as a physical fatigue, affecting mainly endurance, and correction of ID has been shown to improve both fatigue and physical activity [1–8]. In some populations such as heart failure patients, ID, likely through muscle fatigue, is also responsible for a deterioration of the quality of life, a worsening of dyspnea and a worsening of the prognosis of heart failure [9–12]. Interestingly, intravenous iron therapy has also been shown to improve these parameters in heart failure patients, even in the absence of anemia [13–17].

Mechanisms linking ID to muscle fatigue are still poorly understood. The main hypothesis is that ID is responsible for an alteration of mitochondrial muscle metabolism, iron being present in both iron-sulfur centers and in cytochromes of the mitochondrial respiratory chain. Animal studies performed many years ago reported various quantitative and functional anomalies in mitochondrial respiratory chain enzymatic complexes in both skeletal muscles and heart [18–22]. However, the proper role of iron deficiency was difficult to evaluate in these studies, because of a severe anemia systematically observed in their animal models.

We recently developed a mouse model of ID without anemia, in which we assessed the impact of iron deficiency on mitochondrial metabolism in the myocardium [23]. In this model, the mitochondrial respiratory chain complex I activity was decreased in cardiomyocytes, which might explain, at least in part, a significant reduction of left ventricular function and in physical capacity during endurance exercises. However, ID may also impact the mitochondrial metabolism of peripheral skeletal muscle, as recently reported [24,25]. The aim of this study was to assess the impact of ID on the respiratory chain complex I activity of two types of striated muscle fibers in a mouse model of ID without anemia.

2. Materials and Methods

2.1. Animals and Ethics

All experiments were performed in accordance with the guidelines from Directive 2010/63/EU of the European Parliament on the protection of animals used for scientific purposes (laboratory authorization of the laboratory #00577). The protocol was approved by the Ethics Committee in animal experimentation of Pays de la Loire and by the French Ministry of Higher Education and Research (APAFiS #6780).

Male C57BL/6 mice (Janvier, Le Genest St Isle, France) were used for all experiments and all of them were eight-week-old at the start of the experiments. Mice were four per cage, housed in a temperature-controlled room (21 °C) with a 12 h/12 h light-dark cycle. They were fed either with an iron-deficient diet or a normal diet depending on their group and had access to tap water *ad libitum*.

The mouse model of ID without anemia (ID group) was obtained as previously described: on day 1, mice had a 250–300 μL blood withdrawal performed using a retro-orbital collection with a calibrated heparinized capillary tube. Mice were then immediately fed with an iron-depleted diet (C1038 pellets containing 6 mg iron/kg, Genestil SA, Royaucourt, France) for three weeks [23]. As previously shown, mice were considered to have an ID without anemia on the last day of the third week (Day 21) [23]. Control group mice (C group) were fed with a normal diet (M25 pellets containing 150 mg iron/kg, Special Diets Services, France) during the whole study.

Blood withdrawal and sampling were performed under inhaled anesthesia with isoflurane 2% and all efforts were made to minimize suffering. Euthanasia of the animals was made by cervical dislocation, under inhaled anesthesia with isoflurane 2% too, in order to avoid the deleterious effects of CO_2 on mitochondrial metabolism.

2.2. Experimental Design

Each group (ID and C) included eight mice. On day 0 and day 21, mice performed the physical exercises on a Rotarod device. The mice were euthanized on day 21 after the physical exercises and a blood sample, the liver, the spleen and muscle samples of quadriceps and soleus were taken. All muscles were immediately frozen in liquid nitrogen after sampling and stored at −80 °C until analyses.

2.3. Physical Tests

We used a Rotarod device to assess physical capacities of the animals, as already described [23,26]. The Rotarod is a device with a 3 cm diameter cylinder on which mice were individually placed. The cylinder rotated at a progressive acceleration speed followed by a stable speed until the fall of the mouse that made stop the cylinder. The time without falling and the falling speed were automatically recorded (HARotarod software version 1.40) and the calculation of the distance performed by the animals during the total exercise time was done.

The day before the test, the mice were trained to the device by performing an exercise with a constant acceleration speed of 10 to 20 rpm for 180 s followed by a constant speed of

20 rpm. Each time the mice fell, they were immediately put back on the cylinder which restarted its rotation with the same acceleration from 10 to 20 rpm. This training exercise was stopped after 15 min of running on the cylinder in total.

We assessed endurance of the animals over 1 h using two consecutive 30-min tests. On the day of the test (Day 0), the protocol consisted of a first test ("test 20") at a constant acceleration speed of 10 to 20 rpm for 180 s followed by a constant speed of 20 rpm. The test was stopped after 30 min of exercise and was immediately followed by a second test at a constant acceleration speed of 10 to 30 rpm for 180 s followed by a constant speed of 30 rpm. This test was also stopped after 30 min of exercise.

2.4. Hematological Parameters

Hemoglobin concentration (Hb), hematocrit (Ht), mean corpuscular volume (MCV), mean corpuscular hemoglobin (MCH), mean corpuscular hemoglobin concentration (MCHC), reticulocyte count, reticulocyte hemoglobin content (RetHb) and percentage of hypochromic red blood cells (% Hypo RBC) were measured using a hematological automate Sysmex XE-5000 (Sysmex France, Villepinte, France) on the blood samples obtained on Day 21.

2.5. Tissue Iron Content

Splenic and liver iron contents were measured after tissue digestion by trichloroacetic acid, hydrochloric acid and thioglycolic acid, using the iron quantification by the Ferene method on a biochemical automate ARCHITECT c16000 (Abbott France, Rungis, France), as previously described [23,27].

2.6. Mitochondrial Enzymatic Activities

Enzymatic activities were measured in two types of skeletal striated muscle: the quadriceps muscle, which has both oxidative and glycolytic metabolisms, and the soleus, which is mainly oxidative.

Complex I, primarily affected in the heart in our previous study, and complex IV activities were measured at 37 °C with a UVmc2 spectrophotometer (SAFAS, Monaco), according to standard methods [28]. Results were normalized to the citrate synthase activity, an enzyme of the Krebs cycle reflecting the mitochondria content.

Post-nuclear muscle homogenates were prepared at 4 °C. The muscle isolation buffer used was composed of 220 mM mannitol, 75 mM saccharose, 10 mM Tris, 1 mM EGTA adjusted to pH 7.2. The muscle sample was rinsed in the isolation buffer and transferred into a glass tube containing 10 times its weight of the same buffer or 20 times its weight for the soleus. The muscle was homogenized at 1000 rpm with a Potter-Elvehjem PTFE and was centrifuged at 650 g for 20 min. The supernatant was sampled, and the operation was repeated on the pellet. Both supernatants were combined to constitute the post-nuclear muscle homogenate, which was used immediately.

Briefly, NADH ubiquinone reductase (complex I) activity was assayed in KH_2PO_4 buffer (50 mM, pH 7.5), containing 3.75 mg/mL fatty acid-free BSA and 0.1 mM decylubiquinone. 10 mM NADH was added to initiate the reaction. Parallel measurements in presence of rotenone (2.5 µM) were used to determine the background rate. Cytochrome c oxidase (complex IV) activity was assayed in a 92–97% reduced cytochrome c solution. Citrate synthase (CS) activity was assayed in a 0.15 mM DTNB, 0.1% Triton, 0.5 mM oxaloacetic acid and 0.3 mM acetyl-CoA solution. Absorbance changes due to the respective substrate conversions were monitored at 340 nm for complex I, 550 nm for complex IV, and 412 nm for CS. Enzymes activities of complex I and IV, expressed as nmol substrate/min/mg of proteins using the Beer Lambert's law, were then normalized to the CS activity. Reagents were purchased from Sigma-Aldrich (Lyon, France) and NADH from Roche Applied Sciences (Lyon, France).

2.7. Statistical Analysis

Data are reported as medians [interquartiles 25–75%] or numbers (percentages). Categorical and numerical data were compared using the Fisher's exact test and the Mann–Whitney test respectively. All tests were two-tailed and a p-value less than 0.05 was considered significant. Analyses were performed the software JMP (SAS Institute, Inc., Cary, NC, USA).

3. Results

3.1. Description of the Mouse Model of ID

As shown in Figure 1A,B, none of the mice were anemic, with hemoglobin and hematocrit levels greater than 13 g/dL and 40% respectively, without differences between ID and C mice groups. Conversely, the mice of the ID group had hematological signs of ID including a significant decrease in RetHb and a significant increase in % Hypo RBC (Figure 1C–H). Furthermore, tissue iron stores in the spleen and liver were significantly reduced (Figure 1I,J).

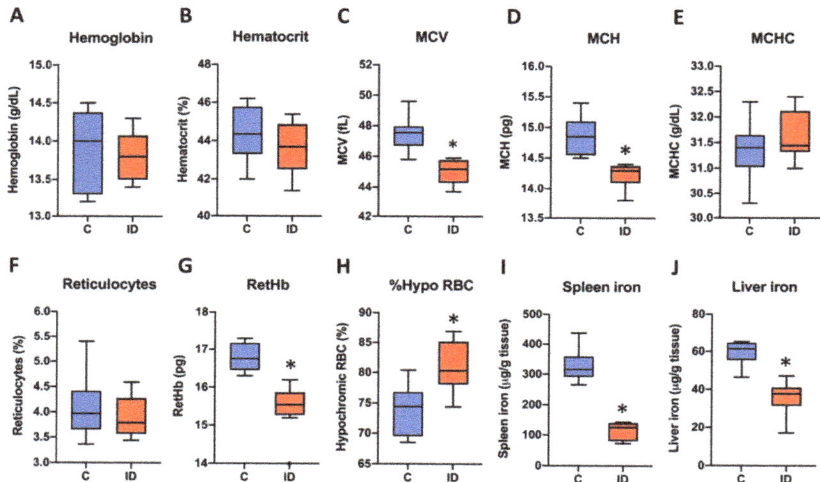

Figure 1. Hematological parameters and iron stores measured on Day 21. (**A**) Hemoglobin concentration; (**B**) hematocrit; (**C**) mean corpuscular volume; (**D**) mean corpuscular hemoglobin; (**E**) mean corpuscular hemoglobin concentration; (**F**) reticulocyte count; (**G**) reticulocyte hemoglobin content; (**H**) percentage of hypochromic red blood cells; (**I**) iron content in the spleen; (**J**) iron content in the liver. C (blue boxes), control group; ID (red boxes), ID group; N = 8 in each group. Box-plots represent medians, interquartile ranges and upper and lower values according to Tukey's method. * $p < 0.05$, significantly different from control group.

3.2. Animal Physical Endurance

The physical capacities of the animals were similar on Day 0 in both groups (Figure 2A–E). On Day 21, ID mice achieved significantly lower maximum times on Rotarod than ID mice on Day 0 and than control mice on Day 0 and Day 21 (Figure 2A–C). The number of falls per test was significantly higher in ID mice on Day 21 than in control mice (Figure 2B–D). Consequently, the distance performed over one hour was significantly reduced in ID than in Control mice (Figure 2E). Mice of the group C had improved their total distance compared to Day 0, although the mice of the group ID had not. We also observed that mice in the ID group had a significantly lower weight at week 3 than Control mice (Figure 2F).

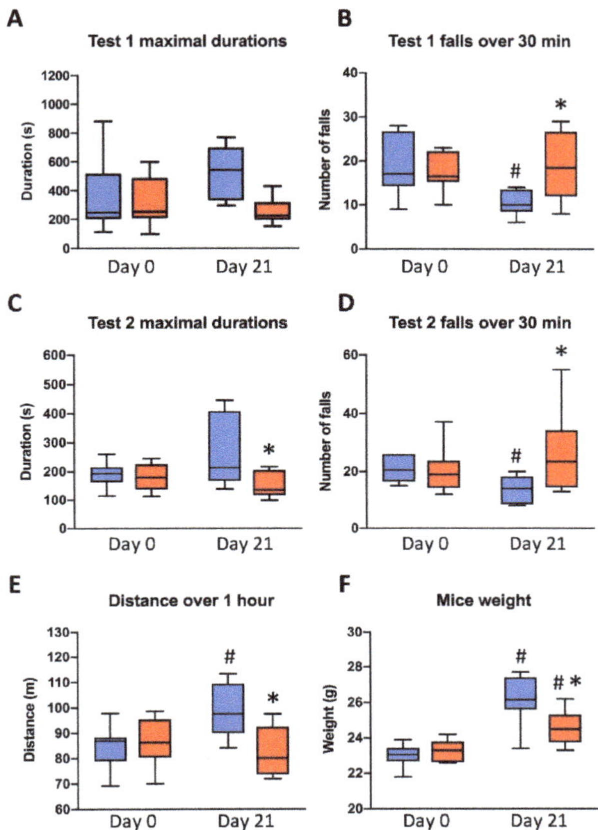

Figure 2. Endurance measured on Rotarod and mice body weights. (**A**) Maximal durations and (**B**) number of falls in test 1 (speed of 10 to 20 rpm for 180 s followed by a constant speed of 20 rpm), (**C**) maximal durations and (**D**) number of falls in test 2 (speed of 10 to 30 rpm for 180 s followed by a constant speed of 30 rpm), (**E**) distance over 1 h performed by mice, and (**F**) weight of mice. C (blue boxes), control group; ID (red boxes), ID group; N = 8 in each group. Box-plots represent medians, interquartile ranges and upper and lower values according to Tukey's method. * $p < 0.05$ compared with control group on the same day; # $p < 0.05$ compared with the same group on Day 0.

3.3. Mitochondrial Enzymatic Activities

The maximal activities of complex I, complex IV and citrate synthase are shown in Figure 3. The activities of complexes I and IV, together with citrate synthase, were all significantly reduced in the soleus muscle of ID mice, suggesting a decrease in mitochondrial content. When the activities of complexes I and IV were normalized to those of citrate synthase, we observed a significant decrease in the activity of complex I in the soleus, without significant differences in activity for complex IV in the soleus and for complexes I and IV in the quadriceps.

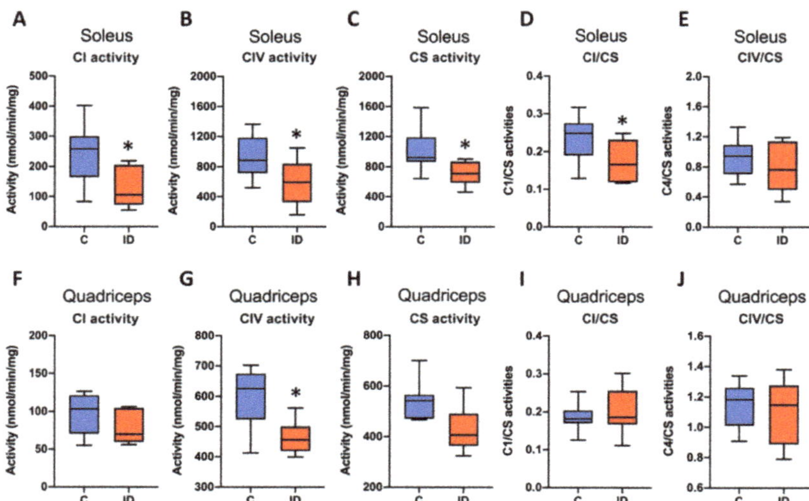

Figure 3. Mitochondrial enzymatic activities. Maximal enzymatic activities of complex I, complex IV, citrate synthase, were measured by spectrophotometry on muscle homogenates of soleus (**A–C**, respectively) and quadriceps (**F–H**, respectively). CI/CS and CIV/CS: specific activities of complexes I and IV normalized to the citrate synthase one (**D,E**: soleus muscle; **I,J**: quadriceps). C (blue boxes), control group; ID (red boxes), ID group; N = 8 in each group. Box-plots represent medians, interquartile ranges and upper and lower values according to Tukey's method. * $p < 0.05$ compared with control group.

4. Discussion

In this study, confirming that ID without anemia is associated with a decrease in endurance capacity, we showed a predominant impact of ID on complex I activity in oxidative skeletal muscle.

The impact of ID, independently of anemia, on physical capacities is difficult to measure. In our mouse model, we used long and submaximal exercises on a Rotarod device to assess endurance. Using this test, we observed a reduction in the distance achieved in one hour of about 15% in ID mice. Overall, previous animal studies found a decrease in both maximum oxygen consumption and endurance capacity [18–21,29]. However, in these studies evaluating physical capacities in ID animals, ID was accompanied by severe anemia, at least at some point of their experiments. Davies et al. showed in a mouse model of ID anemia that anemia mainly alters maximum oxygen consumption [19]. The increase in hemoglobin levels was indeed associated with an early improvement of maximum oxygen consumption after dietary iron repletion, while physical endurance was improved several days later. These results led the authors to suggest that the maximum oxygen consumption, and therefore the capacity to exercise with maximum effort, is mainly due to a reduction of hemoglobin level, whereas endurance seems rather dependent on the oxidative capacities of the muscle, i.e., on mitochondrial metabolism. Conversely, Willis et al. observed in rats fed with an iron-depleted diet a significant decrease in endurance several days before the fall in hemoglobin level and, after an iron dextran injection once the animals were anemic, a very fast improvement (beginning at the 15th hour) of the running time on a treadmill, i.e., before the increase in hemoglobin level [21]. In the model used in the present study, we had previously observed that ID without anemia affected endurance rather than short intense training exercises (e.g., forced swimming exercise) [23]. We confirm here the impact on endurance using a longer test (one hour in total).

In humans, psychic fatigue is a frequent symptom of ID without anemia, which is effectively corrected by iron therapy [1,2]. However, it has also been shown that physical capacities may be improved by iron repletion [30]. The benefit of ID correction, using intravenous iron, has also been demonstrated in heart failure patients [13–17,31]. Interestingly, recent data show that heart failure patients with ID have a greater depletion of phosphocreatine in the gastrocnemius muscle (measured by ^{31}P magnetic resonance spectroscopy), which could partly explain muscle fatigue [24]. Likewise, Charles-Edwards et al. recently observed that iron therapy allows iron-deficient heart failure patients to decrease the regeneration time of skeletal muscle phosphocreatine [25]. Phosphocreatine is necessary during muscle contraction since it makes it possible to give a phosphate to ADP in order to regenerate ATP, and is itself regenerated at rest from ATP supplied by the respiratory chain and glycolysis. Thus, the impairment of physical capacities of these patients could be linked, at least in part, to the impairment of skeletal muscle function due to mitochondrial metabolism dysfunction.

In our mouse model of ID without anemia, we confirmed this impairment of skeletal muscle mitochondrial function in the soleus, but not in the quadriceps. This difference is probably explained by the fact that the soleus muscle is a muscle encompasses a large majority of oxidative type I and type IIA fibers ("red muscle"), while the quadriceps is a mixed muscle containing essentially types II, fast-twitch glycolytic, fibers but a reduced content in type IIA and type I, oxidative, fibers [32]. Type I and IIA fibers, rich in mitochondria, mainly use oxidative capacities of the cell, promoting a higher resistance to fatigue. Conversely, other type II fibers have a lower mitochondria density and have greater glycolytic than oxidative capacities, allowing a significant muscle strength but not a high resistance to fatigue. These particularities probably explain why endurance, which involves fibers with higher oxidative capacities, are mainly affected, in our model and in humans.

We observed in our model that mitochondrial complex I activity was reduced in skeletal muscle in presence of ID, as we had previously observed in the myocardium [23]. This reduction of activity of about 30% seems to be less important than in anemic animals [19], and, while previous studies using anemic animals showed decreased activities of the 4 complexes of the respiratory chain, it seems here to primarily affect the complex I. This result may be explained by the larger number of iron-sulfur centers located in complex I, in comparison with the three other complexes [33]. Iron atoms of iron-sulfur centers, present in complexes I (8–9 iron-sulfur centers), II (3 iron-sulfur centers) and III (1 iron-sulfur center), and iron atoms of heme cytochromes, present in complexes II (1 cytochrome), III (2 cytochromes) and IV (2 cytochromes), have an essential role for the function of the respiratory chain. Indeed, their capacity to change very easily from a ferrous (Fe^{2+}) to a ferric (Fe^{3+}) state facilitates the transfer of electrons required to induce the proton gradient across the inner mitochondrial membrane. Future studies should try to explain the mechanisms leading to the reduction of complex I activity. In our mouse model, we previously observed that this decrease seemed to be linked to a decrease in the overall amount of this complex in the myocardium [23]. The decrease in production of iron-sulfur centers could thus be responsible for a reduction in iron-sulfur centers, leading to a decrease in complex I assembly. However, other mechanisms have been suggested to explain the impact of ID on muscle mitochondrial metabolism, such as an early transition to anaerobic metabolism [34,35], the decrease in the transcription of genes coding for mitochondrial proteins mediated by the IRP/IRE intracellular iron homeostasis system [36], or even mitochondrial morphological alterations [37,38]. Interestingly, we also observed in our study a significant decrease in the activity of citrate synthase in the soleus muscle. The activity of this enzyme, present in the mitochondrial matrix, is commonly used as a reflect of mitochondrial mass [39]. The citrate synthase reduction therefore strongly suggests a decrease in mitochondrial content in soleus muscle, which could be linked to impaired mitochondrial biogenesis or turnover in response to iron ID.

Although these results need to be verified in humans, they open an important avenue for further studies aiming to explore the role of iron in muscle fatigue. A study is underway to evaluate mitochondrial metabolism of cardiomyocytes in cardiac surgery patients, according to their iron status (NCT03541213). Future studies could also focus on skeletal muscle function in patients with or without heart failure. They could allow the widening of iron treatment indications in non-anemic ID patients, as already recommended for heart failure patients [40], especially in situations where recovery of muscle function is required, as in the postoperative period.

This study has some limitations, related in part to the use of an animal model and a limited number of animals. Moreover, we made the choice to use only male mice to create the mouse model of ID without anemia, in order to avoid blood loss linked to the menstrual cycle, that could have been responsible for unwanted anemia. However, both male and female mice will have to be used in future research using this model of ID without anemia, in order to verify whether the impact of iron deficiency is different in female mice. In addition, while the use of a Rotarod device allowed us to evaluate endurance function at a submaximal effort, other functions, such as motivation, coordination or even animal learning are probably involved with the use of this device [41] and may be altered by ID [42,43]. Future human studies will have to use validated tests such as cycloergometer tests, the 6-min walk test, or fatigue and quality of life questionnaires. Finally, due to the small size of soleus muscles, we chose to measure maximal activities of the respiratory chain complexes by spectrophotometry only. Indeed, this method can detect a moderate decrease in an enzymatic complex activity, while oxygraphy only detects this decrease from a higher threshold of inhibition of the complex. However, oxygraphy analyzes will be necessary subsequently to assess the functionality of the whole respiratory chain, in our mouse model and in humans, to further explore the impact of the metabolic changes observed here.

5. Conclusions

In our murine model of ID without anemia, we confirmed that ID negatively affects endurance and is responsible for a decrease in complex I activity in the skeletal striated muscle with a predominant oxidative activity. These results bring new evidence for the rationale of physical fatigue associated to iron deficiency and the potential usefulness of iron therapy to prevent these symptoms.

Supplementary Materials: The following are available online at https://www.mdpi.com/2072-6643/13/4/1056/s1, Raw data S1.

Author Contributions: Conceptualization, E.R., S.L., N.G., V.P. and D.H.; methodology, E.R. and S.L.; validation, P.R., VP., D.H., N.G., F.G. and S.L.; formal analysis, E.R. and N.G.; investigation, E.R., N.G and S.L.; resources, P.R., V.P., F.G. and D.H.; writing—original draft preparation, E.R.; writing—review and editing, S.L., D.H., N.G., V.P., P.R., F.G.; supervision, S.L.; funding acquisition, E.R. All authors have read and agreed to the published version of the manuscript.

Funding: This research was funded by the Société Française d'Anesthésie et de Réanimation (SFAR), grant number Contrat de Recherche 2012.

Institutional Review Board Statement: All experiments were performed in accordance with the guidelines from Directive 2010/63/EU of the European Parliament on the protection of animals used for scientific purposes (laboratory authorization #00577). The protocol was approved by the Ethics Committee in animal experimentation of Pays de la Loire and by the French Ministry of Higher Education and Research (APAFiS #6780).

Data Availability Statement: Data presented in this study are available in Raw data S1.

Acknowledgments: We would like to thank Jennifer Bourreau for her help in animal experiments and Céline Wetterwald for her help in mitochondrial experiments.

Conflicts of Interest: Sigismond Lasocki received consulting fees from Vifor Pharma. All other authors declare no conflict of interest.

References

1.   Krayenbuehl, P.-A.; Battegay, E.; Breymann, C.; Furrer, J.; Schulthess, G. Intravenous Iron for the Treatment of Fatigue in Nonanemic, Premenopausal Women with Low Serum Ferritin Concentration. *Blood* **2011**, *118*, 3222–3227. [CrossRef]

2. Favrat, B.; Balck, K.; Breymann, C.; Hedenus, M.; Keller, T.; Mezzacasa, A.; Gasche, C. Evaluation of a Single Dose of Ferric Carboxymaltose in Fatigued, Iron-Deficient Women–PREFER a Randomized, Placebo-Controlled Study. *PLoS ONE* **2014**, *9*, e94217. [CrossRef]

3. Piednoir, P.; Allou, N.; Driss, F.; Longrois, D.; Philip, I.; Beaumont, C.; Montravers, P.; Lasocki, S. Preoperative Iron Deficiency Increases Transfusion Requirements and Fatigue in Cardiac Surgery Patients: A Prospective Observational Study. *Eur. J. Anaesthesiol.* **2011**, *28*, 796–801. [CrossRef]

4. Lasocki, S.; Chudeau, N.; Papet, T.; Tartiere, D.; Roquilly, A.; Carlier, L.; Mimoz, O.; Seguin, P.; Malledant, Y.; Asehnoune, K.; et al. Prevalence of Iron Deficiency on ICU Discharge and Its Relation with Fatigue: A Multicenter Prospective Study. *Crit. Care Lond. Engl.* **2014**, *18*, 542. [CrossRef]

5. Brownlie, T.; Utermohlen, V.; Hinton, P.S.; Giordano, C.; Haas, J.D. Marginal Iron Deficiency without Anemia Impairs Aerobic Adaptation among Previously Untrained Women. *Am. J. Clin. Nutr.* **2002**, *75*, 734–742. [CrossRef]

6. Zhu, Y.I.; Haas, J.D. Iron Depletion without Anemia and Physical Performance in Young Women. *Am. J. Clin. Nutr.* **1997**, *66*, 334–341. [CrossRef]

7. DellaValle, D.M.; Haas, J.D. Impact of Iron Depletion without Anemia on Performance in Trained Endurance Athletes at the Beginning of a Training Season: A Study of Female Collegiate Rowers. *Int. J. Sport Nutr. Exerc. Metab.* **2011**, *21*, 501–506. [CrossRef] [PubMed]

8. Hinton, P.S.; Giordano, C.; Brownlie, T.; Haas, J.D. Iron Supplementation Improves Endurance after Training in Iron-Depleted, Nonanemic Women. *J. Appl. Physiol. (1985)* **2000**, *88*, 1103–1111. [CrossRef] [PubMed]

9. Jankowska, E.A.; Rozentryt, P.; Witkowska, A.; Nowak, J.; Hartmann, O.; Ponikowska, B.; Borodulin-Nadzieja, L.; Banasiak, W.; Polonski, L.; Filippatos, G.; et al. Iron Deficiency: An Ominous Sign in Patients with Systolic Chronic Heart Failure. *Eur. Heart J.* **2010**, *31*, 1872–1880. [CrossRef] [PubMed]

10. von Haehling, S.; Gremmler, U.; Krumm, M.; Mibach, F.; Schön, N.; Taggeselle, J.; Dahm, J.B.; Angermann, C.E. Prevalence and Clinical Impact of Iron Deficiency and Anaemia among Outpatients with Chronic Heart Failure: The PrEP Registry. *Clin. Res. Cardiol. Off. J. Ger. Card. Soc.* **2017**, *106*, 436–443. [CrossRef]

11. Klip, I.T.; Comin-Colet, J.; Voors, A.A.; Ponikowski, P.; Enjuanes, C.; Banasiak, W.; Lok, D.J.; Rosentryt, P.; Torrens, A.; Polonski, L.; et al. Iron Deficiency in Chronic Heart Failure: An International Pooled Analysis. *Am. Heart J.* **2013**, *165*, 575–582.e3. [CrossRef] [PubMed]

12. Bekfani, T.; Pellicori, P.; Morris, D.; Ebner, N.; Valentova, M.; Sandek, A.; Doehner, W.; Cleland, J.G.; Lainscak, M.; Schulze, P.C.; et al. Iron Deficiency in Patients with Heart Failure with Preserved Ejection Fraction and Its Association with Reduced Exercise Capacity, Muscle Strength and Quality of Life. *Clin. Res. Cardiol. Off. J. Ger. Card. Soc.* **2019**, *108*, 203–211. [CrossRef]

13. Toblli, J.E.; Lombraña, A.; Duarte, P.; Di Gennaro, F. Intravenous Iron Reduces NT-pro-Brain Natriuretic Peptide in Anemic Patients with Chronic Heart Failure and Renal Insufficiency. *J. Am. Coll. Cardiol.* **2007**, *50*, 1657–1665. [CrossRef]

14. Okonko, D.O.; Grzeslo, A.; Witkowski, T.; Mandal, A.K.J.; Slater, R.M.; Roughton, M.; Foldes, G.; Thum, T.; Majda, J.; Banasiak, W.; et al. Effect of Intravenous Iron Sucrose on Exercise Tolerance in Anemic and Nonanemic Patients with Symptomatic Chronic Heart Failure and Iron Deficiency FERRIC-HF: A Randomized, Controlled, Observer-Blinded Trial. *J. Am. Coll. Cardiol.* **2008**, *51*, 103–112. [CrossRef]

15. Anker, S.D.; Comin Colet, J.; Filippatos, G.; Willenheimer, R.; Dickstein, K.; Drexler, H.; Lüscher, T.F.; Bart, B.; Banasiak, W.; Niegowska, J.; et al. Ferric Carboxymaltose in Patients with Heart Failure and Iron Deficiency. *N. Engl. J. Med.* **2009**, *361*, 2436–2448. [CrossRef] [PubMed]

16. Filippatos, G.; Farmakis, D.; Colet, J.C.; Dickstein, K.; Lüscher, T.F.; Willenheimer, R.; Parissis, J.; Gaudesius, G.; Mori, C.; von Eisenhart Rothe, B.; et al. Intravenous Ferric Carboxymaltose in Iron-Deficient Chronic Heart Failure Patients with and without Anaemia: A Subanalysis of the FAIR-HF Trial. *Eur. J. Heart Fail.* **2013**, *15*, 1267–1276. [CrossRef] [PubMed]

17. van Veldhuisen, D.J.; Ponikowski, P.; van der Meer, P.; Metra, M.; Böhm, M.; Doletsky, A.; Voors, A.A.; Macdougall, I.C.; Anker, S.D.; Roubert, B.; et al. Effect of Ferric Carboxymaltose on Exercise Capacity in Patients With Chronic Heart Failure and Iron Deficiency. *Circulation* **2017**, *136*, 1374–1383. [CrossRef] [PubMed]

18. Finch, C.A.; Miller, L.R.; Inamdar, A.R.; Person, R.; Seiler, K.; Mackler, B. Iron Deficiency in the Rat. Physiological and Biochemical Studies of Muscle Dysfunction. *J. Clin. Investig.* **1976**, *58*, 447–453. [CrossRef]

19. Davies, K.J.; Maguire, J.J.; Brooks, G.A.; Dallman, P.R.; Packer, L. Muscle Mitochondrial Bioenergetics, Oxygen Supply, and Work Capacity during Dietary Iron Deficiency and Repletion. *Am. J. Physiol.* **1982**, *242*, E418–E427. [CrossRef] [PubMed]

20. Davies, K.J.; Donovan, C.M.; Refino, C.J.; Brooks, G.A.; Packer, L.; Dallman, P.R. Distinguishing Effects of Anemia and Muscle Iron Deficiency on Exercise Bioenergetics in the Rat. *Am. J. Physiol.* **1984**, *246*, E535–E543. [CrossRef]

21. Willis, W.T.; Brooks, G.A.; Henderson, S.A.; Dallman, P.R. Effects of Iron Deficiency and Training on Mitochondrial Enzymes in Skeletal Muscle. *J. Appl. Physiol. (1985)* **1987**, *62*, 2442–2446. [CrossRef]

22. Blayney, L.; Bailey-Wood, R.; Jacobs, A.; Henderson, A.; Muir, J. The Effects of Iron Deficiency on the Respiratory Function and Cytochrome Content of Rat Heart Mitochondria. *Circ. Res.* **1976**, *39*, 744–748. [CrossRef] [PubMed]

23. Rineau, E.; Gaillard, T.; Gueguen, N.; Procaccio, V.; Henrion, D.; Prunier, F.; Lasocki, S. Iron Deficiency without Anemia Is Responsible for Decreased Left Ventricular Function and Reduced Mitochondrial Complex I Activity in a Mouse Model. *Int. J. Cardiol.* **2018**, *266*, 206–212. [CrossRef] [PubMed]

24. Melenovsky, V.; Hlavata, K.; Sedivy, P.; Dezortova, M.; Borlaug, B.A.; Petrak, J.; Kautzner, J.; Hajek, M. Skeletal Muscle Abnormalities and Iron Deficiency in Chronic Heart FailureAn Exercise 31P Magnetic Resonance Spectroscopy Study of Calf Muscle. *Circ. Heart Fail.* **2018**, *11*, e004800. [CrossRef]

25. Charles-Edwards, G.; Amaral, N.; Sleigh, A.; Ayis, S.; Catibog, N.; McDonagh, T.; Monaghan, M.; Amin-Youssef, G.; Kemp, G.J.; Shah, A.M.; et al. Effect of Iron Isomaltoside on Skeletal Muscle Energetics in Patients With Chronic Heart Failure and Iron Deficiency. *Circulation* **2019**, *139*, 2386–2398. [CrossRef] [PubMed]

26. Burnes, L.A.; Kolker, S.J.; Danielson, J.F.; Walder, R.Y.; Sluka, K.A. Enhanced Muscle Fatigue Occurs in Male but Not Female ASIC3-/- Mice. *Am. J. Physiol. Regul. Integr. Comp. Physiol.* **2008**, *294*, R1347–R1355. [CrossRef]

27. Lasocki, S.; Millot, S.; Andrieu, V.; Lettéron, P.; Pilard, N.; Muzeau, F.; Thibaudeau, O.; Montravers, P.; Beaumont, C. Phlebotomies or Erythropoietin Injections Allow Mobilization of Iron Stores in a Mouse Model Mimicking Intensive Care Anemia. *Crit. Care Med.* **2008**, *36*, 2388–2394. [CrossRef]

28. Medja, F.; Allouche, S.; Frachon, P.; Jardel, C.; Malgat, M.; Mousson de Camaret, B.; Slama, A.; Lunardi, J.; Mazat, J.P.; Lombès, A. Development and Implementation of Standardized Respiratory Chain Spectrophotometric Assays for Clinical Diagnosis. *Mitochondrion* **2009**, *9*, 331–339. [CrossRef]

29. McLane, J.A.; Fell, R.D.; McKay, R.H.; Winder, W.W.; Brown, E.B.; Holloszy, J.O. Physiological and Biochemical Effects of Iron Deficiency on Rat Skeletal Muscle. *Am. J. Physiol.* **1981**, *241*, C47–C54. [CrossRef] [PubMed]

30. Brutsaert, T.D.; Hernandez-Cordero, S.; Rivera, J.; Viola, T.; Hughes, G.; Haas, J.D. Iron Supplementation Improves Progressive Fatigue Resistance during Dynamic Knee Extensor Exercise in Iron-Depleted, Nonanemic Women. *Am. J. Clin. Nutr.* **2003**, *77*, 441–448. [CrossRef]

31. Ponikowski, P.; van Veldhuisen, D.J.; Comin-Colet, J.; Ertl, G.; Komajda, M.; Mareev, V.; McDonagh, T.; Parkhomenko, A.; Tavazzi, L.; Levesque, V.; et al. Beneficial Effects of Long-Term Intravenous Iron Therapy with Ferric Carboxymaltose in Patients with Symptomatic Heart Failure and Iron Deficiency†. *Eur. Heart J.* **2015**, *36*, 657–668. [CrossRef] [PubMed]

32. Augusto, V.; Padovani, C.; Eduardo, G.; Campos, R. Skeletal Muscle Fiber Types in C57BL6J Mice. *J. Morphol. Sci.* **2004**, *21*, 89–94.

33. Hinchliffe, P.; Sazanov, L.A. Organization of Iron-Sulfur Clusters in Respiratory Complex I. *Science* **2005**, *309*, 771–774. [CrossRef]

34. Klempa, K.L.; Willis, W.T.; Chengson, R.; Dallman, P.R.; Brooks, G.A. Iron Deficiency Decreases Gluconeogenesis in Isolated Rat Hepatocytes. *J. Appl. Physiol. (1985)* **1989**, *67*, 1868–1872. [CrossRef] [PubMed]

35. Henderson, S.A.; Dallman, P.R.; Brooks, G.A. Glucose Turnover and Oxidation Are Increased in the Iron-Deficient Anemic Rat. *Am. J. Physiol.* **1986**, *250*, E414–E421. [CrossRef]

36. Eisenstein, R.S. Iron Regulatory Proteins and the Molecular Control of Mammalian Iron Metabolism. *Annu. Rev. Nutr.* **2000**, *20*, 627–662. [CrossRef]

37. Cartier, L.J.; Ohira, Y.; Chen, M.; Cuddihee, R.W.; Holloszy, J.O. Perturbation of Mitochondrial Composition in Muscle by Iron Deficiency. Implications Regarding Regulation of Mitochondrial Assembly. *J. Biol. Chem.* **1986**, *261*, 13827–13832. [CrossRef]

38. Dong, F.; Zhang, X.; Culver, B.; Chew, H.G.; Kelley, R.O.; Ren, J. Dietary Iron Deficiency Induces Ventricular Dilation, Mitochondrial Ultrastructural Aberrations and Cytochrome c Release: Involvement of Nitric Oxide Synthase and Protein Tyrosine Nitration. *Clin. Sci.* **2005**, *109*, 277–286. [CrossRef]

39. Larsen, S.; Nielsen, J.; Hansen, C.N.; Nielsen, L.B.; Wibrand, F.; Stride, N.; Schroder, H.D.; Boushel, R.; Helge, J.W.; Dela, F.; et al. Biomarkers of Mitochondrial Content in Skeletal Muscle of Healthy Young Human Subjects. *J. Physiol.* **2012**, *590*, 3349–3360. [CrossRef]

40. Ponikowski, P.; Voors, A.A.; Anker, S.D.; Bueno, H.; Cleland, J.G.F.; Coats, A.J.S.; Falk, V.; González-Juanatey, J.R.; Harjola, V.-P.; Jankowska, E.A.; et al. 2016 ESC Guidelines for the Diagnosis and Treatment of Acute and Chronic Heart Failure: The Task Force for the Diagnosis and Treatment of Acute and Chronic Heart Failure of the European Society of Cardiology (ESC). Developed with the Special Contribution of the Heart Failure Association (HFA) of the ESC. *Eur. J. Heart Fail.* **2016**, *18*, 891–975. [CrossRef]

41. Galante, M.; Jani, H.; Vanes, L.; Daniel, H.; Fisher, E.M.C.; Tybulewicz, V.L.J.; Bliss, T.V.P.; Morice, E. Impairments in Motor Coordination without Major Changes in Cerebellar Plasticity in the Tc1 Mouse Model of Down Syndrome. *Hum. Mol. Genet.* **2009**, *18*, 1449–1463. [CrossRef] [PubMed]

42. Li, R.; Chen, X.; Yan, H.; Deurenberg, P.; Garby, L.; Hautvast, J.G. Functional Consequences of Iron Supplementation in Iron-Deficient Female Cotton Mill Workers in Beijing, China. *Am. J. Clin. Nutr.* **1994**, *59*, 908–913. [CrossRef] [PubMed]

43. Crouter, S.E.; DellaValle, D.M.; Haas, J.D. Relationship between Physical Activity, Physical Performance, and Iron Status in Adult Women. *Appl. Physiol. Nutr. Metab. Physiol. Appl. Nutr. Metab.* **2012**, *37*, 697–705. [CrossRef] [PubMed]

Article

Krill Oil Supplementation Reduces Exacerbated Hepatic Steatosis Induced by Thermoneutral Housing in Mice with Diet-Induced Obesity

Gabriella Sistilli [1,2], Veronika Kalendova [1,2], Tomas Cajka [1], Illaria Irodenko [1,2], Kristina Bardova [1], Marina Oseeva [1,2], Petr Zacek [3], Petra Kroupova [1], Olga Horakova [1], Karoline Lackner [4], Amalia Gastaldelli [5], Ondrej Kuda [1], Jan Kopecky [1] and Martin Rossmeisl [1,*]

1   Institute of Physiology of the Czech Academy of Sciences, Videnska 1083, 14220 Prague 4, Czech Republic; gabriella.sistilli@fgu.cas.cz (G.S.); veronika.kalendova@fgu.cas.cz (V.K.); tomas.cajka@fgu.cas.cz (T.C.); illaria.irodenko@fgu.cas.cz (I.I.); kristina.bardova@fgu.cas.cz (K.B.); marina.oseeva@fgu.cas.cz (M.O.); petra.kroupova@fgu.cas.cz (P.K.); olga.horakova@fgu.cas.cz (O.H.); ondrej.kuda@fgu.cas.cz (O.K.); jan.kopecky@fgu.cas.cz (J.K.)
2   Department of Physiology, Faculty of Science, Charles University, Vinicna 7, 12844 Prague 2, Czech Republic
3   Proteomics Core Facility, Faculty of Science, Charles University, Division BIOCEV, Prumyslova 595, 25250 Vestec, Czech Republic; zacek@natur.cuni.cz
4   Institute of Pathology, Medical University of Graz, Neue Stiftingtalstraße 6, 8010 Graz, Austria; karoline.lackner@medunigraz.at
5   Cardiometabolic Risk Unit, Institute of Clinical Physiology, National Research Council, Via Moruzzi 1, 56100 Pisa, Italy; amalia@ifc.cnr.it
*   Correspondence: martin.rossmeisl@fgu.cas.cz; Tel.: +420-2-4106-3706

check for **updates**

Citation: Sistilli, G.; Kalendova, V.; Cajka, T.; Irodenko, I.; Bardova, K.; Oseeva, M.; Zacek, P.; Kroupova, P.; Horakova, O.; Lackner, K.; et al. Krill Oil Supplementation Reduces Exacerbated Hepatic Steatosis Induced by Thermoneutral Housing in Mice with Diet-Induced Obesity. *Nutrients* 2021, 13, 437. https://doi.org/10.3390/nu13020437

Academic Editors: Arie Nieuwenhuizen and Lindsay Brown

Received: 4 December 2020
Accepted: 26 January 2021
Published: 29 January 2021

Publisher's Note: MDPI stays neutral with regard to jurisdictional claims in published maps and institutional affiliations.

Abstract: Preclinical evidence suggests that n-3 fatty acids EPA and DHA (Omega-3) supplemented as phospholipids (PLs) may be more effective than triacylglycerols (TAGs) in reducing hepatic steatosis. To further test the ability of Omega-3 PLs to alleviate liver steatosis, we used a model of exacerbated non-alcoholic fatty liver disease based on high-fat feeding at thermoneutral temperature. Male C57BL/6N mice were fed for 24 weeks a lard-based diet given either alone (LHF) or supplemented with Omega-3 (30 mg/g diet) as PLs (krill oil; ω3PL) or TAGs (Epax 3000TG concentrate; ω3TG), which had a similar total content of EPA and DHA and their ratio. Substantial levels of TAG accumulation (~250 mg/g) but relatively low inflammation/fibrosis levels were achieved in the livers of control LHF mice. Liver steatosis was reduced by >40% in the ω3PL but not ω3TG group, and plasma ALT levels were markedly reduced (by 68%) in ω3PL mice as well. Krill oil administration also improved hepatic insulin sensitivity, and its effects were associated with high plasma adiponectin levels (150% of LHF mice) along with superior bioavailability of EPA, increased content of alkaloids stachydrine and trigonelline, suppression of lipogenic gene expression, and decreased diacylglycerol levels in the liver. This study reveals that in addition to Omega-3 PLs, other constituents of krill oil, such as alkaloids, may contribute to its strong antisteatotic effects in the liver.

Keywords: NAFLD; obesity; omega-3; krill oil; phospholipids; high-fat diet; C57BL/6N mice; thermoneutral temperature

1. Introduction

Obesity is frequently associated with non-alcoholic fatty liver disease (NAFLD), a spectrum of conditions ranging from increased intrahepatic accumulation of triacylglycerols (TAGs; i.e., fatty liver or hepatic steatosis) to steatohepatitis (NASH) and end-stage liver disease [1]. Prevalence of hepatic steatosis and NASH in extremely obese subjects may reach up to 85% and 40%, respectively [2,3], while the presence of metabolic syndrome is associated with a potentially progressive, severe liver disease [4,5]. NAFLD is a serious public health problem [6] for which currently no approved drug therapy exists [7].

Dietary fatty acids (FAs) can differentially affect the body's ability to store lipids in certain fat depots as well as in extra-adipose tissues [8]. In humans, overeating saturated FAs (SFAs) promoted hepatic and visceral fat storage [9–11]. Differential effects of various types of FAs are also observed in the case of regulation of inflammatory responses; thus SFAs and polyunsaturated FAs (PUFAs) of n-6 series are more pro-inflammatory, while PUFAs of n-3 series (omega-3 PUFAs) such as docosahexaenoic acid (DHA; 22:6n-3) and eicosapentaenoic acid (EPA; 20:5n-3) exert anti-inflammatory and hypolipidemic effects ([12–14], and reviewed in [15–17]). At the same time, omega-3 PUFA supplementation may reduce de novo lipogenesis (DNL) and increase FA oxidation in the liver [18], with the transcription factor peroxisome proliferator-activated receptor (PPAR)α playing a crucial role in the latter effect [19]. For these reasons, omega-3 PUFA supplements could be effective in preventing and treating NAFLD [20]. Indeed, in NAFLD patients treated with EPA and DHA as ethyl esters, a decrease in the percentage of liver fat was linearly correlated with the amount of omega-3 PUFAs taken [21]; however, no improvement in markers of liver function/injury or the fibrosis scores was detected. Similar results were obtained in subjects with non-cirrhotic NASH treated with omega-3 PUFAs as TAGs [22], in which a decrease in liver fat but no improvements in histological activity were observed. In general, omega-3 PUFAs administered in the form of TAGs or ethyl esters have been shown to partially limit hepatic steatosis in some studies [23].

Omega-3 PUFAs also alleviated hepatic steatosis in various rodent models of obesity (e.g., [14,24–26], and reviewed in [27]). Interestingly, the efficacy of omega-3 PUFAs may depend on the lipid form of their supplementation. For instance, compared to their TAG form, omega-3 PUFAs administered via phosphatidylcholine-rich phospholipids (PLs), either in the form of krill oil extracted from the Antarctic krill *Euphausia superba* [28] or as an extract of herring meal [26], had stronger effects in reducing the TAG content in the liver of rodents with genetically- or diet-induced obesity [25,26,29]. Moreover, in mice fed a corn oil-based high-fat diet, a significant reduction in hepatic TAGs was achieved only by administration of marine PLs containing omega-3 PUFAs and not soybean-derived phosphatidylcholine that did not contain EPA or DHA [30]. The higher efficacy of omega-3 PUFA-containing PLs in reducing hepatic steatosis could be related to the improved bioavailability of omega-3 PUFAs, in particular EPA and docosapentaenoic acid (22:5n-3), both in plasma and in target organs ([26,29,31] and reviewed in [32]). At the same time, supplementation of omega-3 PUFAs as PLs led to a stronger downregulation of liver gene expression in the DNL pathway [29,33,34] and significantly reduced activities of the corresponding lipogenic enzymes as well as of the mitochondrial citrate carrier [35]. However, despite its strong effects on TAG accumulation in the liver, it is not clear whether administration of omega-3 PUFAs in the form of PLs is able to affect advanced stages of NAFLD such as NASH and fibrosis, which remain unaffected in response to more traditional forms of omega-3 PUFAs such as TAGs or ethyl esters (see above).

Recently, a mouse model of obesity-associated exacerbated NAFLD based on the administration of a lard-based high-fat diet in a thermoneutral environment was introduced [36]. This experimental model is characterized by lower stress-driven production of corticosterone, augmented mouse pro-inflammatory immune responses and markedly exacerbated high-fat diet-induced NAFLD pathogenesis, which should recapitulate the severe end of the disease spectrum in humans. Thus, in the present study, the above experimental conditions were used to examine whether omega-3 PUFAs supplemented as PLs via krill oil could beneficially affect NAFLD-related phenotypes and hepatic insulin sensitivity, and what is the potential mechanism of action. At the same time, for comparison, other mice were administered omega-3 PUFAs in the form of a TAG-based concentrate, which was similar to krill oil in terms of the amount and ratio of EPA and DHA, thus representing the group receiving omega-3 PUFAs in one of the traditional lipid forms used for this purpose.

2. Materials and Methods

2.1. Animals and Diets

Male C57BL/6N mice (Charles River Laboratories, Sulzfeld, Germany) were obtained at the age of ~10 weeks. After arrival, mice were individually housed in cages and maintained at ~22 °C on a 12-h light/dark cycle (light from 6:00 a.m.), with ad libitum access to water and a standard diet (Chow; ~14 kJ/g; fat content ~3.6% (*w/w*); Rat/Mouse-Maintenance extrudate; ssniff Spezialdiäten GmbH, Soest, Germany). After one week of adaptation, the animals were transferred to a thermoneutral environment (~30 °C) and fed the following experimental diets: (i) a lard-based high-fat diet (LHF diet; ~21 kJ/g; fat content ~35% (*w/w*); product "DIO-60 kJ% fat (Lard)", Cat. No. E15742-34; ssniff Spezialdiäten GmbH, Soest, Germany), (ii) a LHF-based diet supplemented with omega-3 PUFA-containing PLs (ω3PL diet), using krill oil (Rimfrost Sublime; EPA ~13%, DHA ~8%; Rimfrost AS, Ålesund, Norway), and (iii) a LHF-based diet supplemented with omega-3 PUFAs in the form of re-esterified TAGs (ω3TG diet), using the product Epax 3000 TG (EPA ~18%, DHA ~11%; Epax Norway AS, Ålesund, Norway). Experimental diets were prepared at the ssniff facility in Germany. The total content of EPA and DHA in both supplemented diets was ~30 mg/g diet. For details on macronutrient and FA composition of the experimental diets, see Supplementary Materials Tables S1 and S2, respectively.

2.2. Experimental Setup

The experimental setup is shown in Figure 1. After moving to a thermoneutral environment, the mice were divided into four groups with the same average weight ($n = 8$; see Figure 1A). One group was fed the LHF diet for 24 weeks, as was the group fed the ω3PL diet from the beginning of the experiment (i.e., "preventive" approach). However, the other two groups first received the LHF diet for 8 weeks, and only then was this diet replaced with omega-3 PUFA-supplemented diets (ω3TG or ω3PL) administered for the remaining 16 weeks (i.e., "reverse" approach; marked with the letter "R" at the end of the group name). The total duration of all dietary interventions was therefore 24 weeks. Chow-fed mice served as lean controls. Body weight was recorded weekly and a fresh ration of diet was administered every other day. The calculation of cumulative energy intake was based on weekly measurements of food consumption over 24 h. Fasting plasma insulin and blood glucose levels were measured at week 21. Mice were killed by cervical dislocation under diethyl ether anesthesia between 9:00 a.m. and 11:00 a.m. Truncal blood was collected into tubes containing EDTA for plasma isolation, liver and white adipose tissue (WAT) samples from the epididymal, mesenteric and subcutaneous (dorso-lumbar) fat depots were dissected, weighed, and adiposity index was calculated as the sum of the weights of all analyzed WAT depots divided by body weight. Liver and epididymal WAT samples were snap-frozen in liquid nitrogen and stored at −80 °C for subsequent analyses, while one aliquot of tissue was used for histological evaluation. In a separate study using mice from the Chow, LHF and ω3PL groups (see Figure 1B), pyruvate tolerance, hepatic production of TAGs contained in very low-density lipoproteins (VLDL) and insulin sensitivity were determined at weeks 21, 23, and 24, respectively. Animal experiments were approved by the Institutional Animal Care and Use Committee and the Committee for Animal Protection of the Ministry of Agriculture of the Czech Republic (Approval Number: 81/2016).

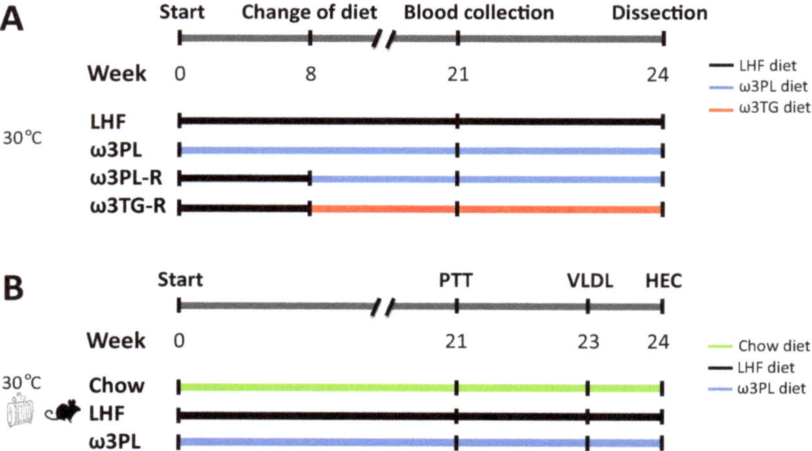

Figure 1. Overview of the experimental setup. (**A**) Four groups of mice (*n* = 8) housed in a thermoneutral environment (~30 °C) were used: (i) the control LHF group, which was fed a lard-based high-fat diet (i.e., LHF diet) for 24 weeks; (ii) ω3PL group fed a LHF-based diet supplemented with omega-3 PUFAs as PLs in the form of krill oil (i.e., ω3PL diet) for the duration of the experiment (i.e., "preventive" approach); (iii) ω3PL-R group fed the LHF diet for the first eight weeks and then from the ninth week on the ω3PL diet until the end of the experiment (i.e., "reverse" approach; marked with the letter "R" at the end of the group name); and (iv) ω3TG-R group fed the LHF diet for the first 8 weeks and then from the ninth week on the LHF-based diet supplemented with omega-3 PUFAs in the form of a concentrate of re-esterified TAGs (i.e., ω3TG diet) until the end of the experiment. (**B**) Three groups of mice (*n* = 8) housed in a thermoneutral environment (~30 °C) were used: (i) Chow group, in which mice were fed a standard low-fat diet and served as lean controls; (ii) the control LHF group, which was fed a lard-based high-fat diet (i.e., LHF diet) for 24 weeks; and (iii) ω3PL group fed the ω3PL diet for the duration of the experiment. Further details in Section 2.2. PTT, pyruvate tolerance test; VLDL, liver VLDL-TAGs secretion test; HEC, hyperinsulinemic-euglycemic clamp.

2.3. Pyruvate Tolerance Test

The level of gluconeogenesis was estimated using pyruvate tolerance test. Mice fasted overnight (~14 h) were injected i.p. with pyruvate (1.5 mg/g body weight) and blood glucose levels were measured using glucometers Contour Plus (Bayer, Leverkusen, Germany) at time 0 (i.e., before injection), and then 15, 30, 60, 120, and 180 min after injection. The response to pyruvate administration was quantified as area under the glucose curve (AUC).

2.4. Light Microscopy and Immunohistochemical Analysis

Liver and epididymal WAT samples were fixed in 4% formaldehyde, embedded in paraffin, and sections of 5 μm thickness were stained using hematoxylin-eosin. The NAFLD histological scoring system [37] was used to assess the effect of omega-3 PUFAs administration on NAFLD progression. In epididymal WAT, macrophage marker MAC-2/galectin-3 was detected using specific antibodies (Cedarlane Laboratories; Burlington, NC, USA; 1:4000 dilution) and the number of crown-like structures (CLS) counted as before [38]. Morphometric analysis of WAT was performed using the imaging software NIS-Elements AR3.0 (Laboratory Imaging, Prague, Czech Republic).

2.5. Hepatic Production of VLDL-TAGs

The procedure was the same as before [39]. After an overnight fast (~16 h), mice were injected i.p. with a solution of 15% Tyloxapol (Triton WR-1339; Sigma-Aldrich; Prague, Czech Republic; dissolved in 0.9% saline) at a dose of 500 mg per kg body weight and blood was collected from the tail vein under basal conditions and 2, 4, and 6 h after Tyloxapol injection. Plasma TAG concentrations were measured at 500 nm using the Triglycerides kit from Erba Lachema (Brno, Czech Republic) and the Sunrise microplate reader (Tecan Group, Männedorf, Switzerland).

2.6. Insulin Sensitivity Measured by Hyperinsulinemic-Euglycemic Clamp

Hyperinsulinemic-euglycemic clamp was performed in awake mice as before [14,29]. Briefly, a week before the end of the study, a permanent catheter was inserted into the *v. jugularis*. After a postoperative period of 4–7 days, mice were fasted for 6 h (6:00 a.m.–12:00 p.m.) and then infused with insulin Actrapid (Novo Nordisk) and D-[3-^{3}H]glucose (Perkin Elmer, Boston, MA, USA) at a constant rate of 4.8 mU/min per kg body weight and 0.26 µCi/min, respectively. Euglycemia (~5.5 mmol/L) was maintained by periodical adjusting the variable infusion of glucose solution (30% for lean animals, 15% for obese animals), while blood glucose levels were regularly monitored using glucometers (see Section 2.3). Blood samples taken every 10 min during the last hour of the 3-h infusion period were used to analyze specific D-[3-^{3}H]-glucose activity.

2.7. Metabolites and Hormones

Plasma levels of lipid metabolites (i.e., TAGs, total cholesterol, non-esterified fatty acids), as well as aspartate transaminase (AST) and alanine aminotransferase (ALT), were measured using the appropriate assays from Roche or Wako (for the measurement of non-esterified fatty acids) and a Clinical Chemistry analyzer Roche/Hitachi 902 (Roche Diagnostics; Basel, Switzerland). Plasma levels of insulin were quantified using xMAP technology and MILLIPLEX MAP Mouse Metabolic Hormone Magnetic Bead Panel (MMHMAG-44K; Merck-Millipore; Burlington, MA, USA). Plasma levels of total adiponectin were measured by Mouse Adiponectin ELISA kit (EZMADP-60K; Sigma-Aldrich). Fasting plasma insulin and blood glucose levels were used to quantify Homeostatic Model Assessment of Insulin Resistance (HOMA-IR), using the following formula: fasting plasma insulin (mU/L) × fasting plasma glucose (mmol/L)/22.5.

2.8. TAG Content in the Liver

Approximately 50 mg of tissue was dissolved in 150 µL 3M KOH (dissolved in 65% ethanol) at 70 °C for 2 h. The resulting homogenate was diluted 10× in redistilled water and the TAG content was measured (see Section 2.5) and the results were related to tissue weight.

2.9. Gene Expression Analysis

Gene expression was analyzed in the liver (stored in RNA later; Ambion, Austin, TX, USA) by real-time quantitative PCR as before [29,40]. Transcript levels were normalized to the expression level of a housekeeping gene for 18S ribosomal RNA (*Rn18s*). Gene names and sequences of the oligonucleotide primers are listed in Table S3.

2.10. Composition of FAs in Experimental Diets and Liver

Total lipids were extracted from aliquots of experimental diets (100 mg) by two-step extraction using hexane and a mixture of methanol and dichloromethane (see Supplementary Materials for details). The methyl tert-butyl ether (MTBE)-based extraction of total lipids from the liver (50 mg) was performed as before [31], and neutral and polar lipid fractions were obtained by using SPE Columns (Discovery). Trans-esterification of extracted lipids, FAs methyl esters (FAME) extraction and their analysis using comprehensive two-dimensional gas chromatography with mass detection (Pegasus 4D, LECO, USA) was performed as before [41].

2.11. LC-MS Analysis of Liver Samples

Metabolomic and lipidomic profiling of liver samples was conducted using a combined targeted and untargeted workflow for the lipidome, metabolome, and exposome analysis (LIMeX) [42,43]. Extraction was carried out using a biphasic solvent system of cold methanol, MTBE, and 10% methanol. Four different LC-MS platforms were used for metabolomic and lipidomic profiling: (i) lipidomics of complex lipids in positive ion mode, (ii) lipidomics of complex lipids in negative ion mode, (iii) metabolomics of polar metabolites in positive ion mode, and (iv) metabolomics of polar metabolites in negative ion mode. Details of sample preparation, LC-MS conditions, raw data processing and curation, and list of annotated complex lipids and polar metabolites are in Supplementary Materials.

2.12. Data Processing and Statistics

Results are means $\pm$ SEM. To compare the groups fed experimental LHF-based diets, One Way ANOVA (for normally distributed data sets) or Kruskal–Wallis test (non-normally distributed data sets) followed by Student–Newman–Keuls post-hoc test was used (Sigma-Stat 3.5 software; Systat Software Inc., San Jose, CA, USA). Differences were considered significant when $p \leq 0.05$. Pearson's correlation coefficient (r) was calculated to measure the strength of the association between the two variables. Multivariate analysis was performed using partial least-squares discriminant analysis (PLS-DA) using MetaboAnalyst 4.0 [44]. Statistical models were created for metabolomic and lipidomic sets separately after logarithmic transformation (base 10) and Pareto scaling. Exported variable importance in projection (VIP) scores were used for evaluation.

The number of animals required to evaluate the effect of krill oil supplementation on liver steatosis was based on a previous publication [29], namely the difference in liver TAGs between control mice fed a corn oil-based high-fat diet (i.e., cHF diet) and mice fed a cHF-based diet containing krill oil (i.e., the ω3PL-H diet), supplemented at the same dose as in the current study. Thus, the minimal sample size of 6 animals per group was calculated using G*Power software (power 0.95, α = 0.05; see [45]).

3. Results

3.1. Basic Parameters of Energy Balance, Adiposity, as well as Lipid and Glucose Homeostasis

Table 1 and Figure 2 show the general characteristics of obese mice from the control group (LHF) and the three intervention groups (ω3PL, ω3PL-R and ω3TG-R; description of groups—see Section 2.2 and Figure 1) that differed in the lipid form and/or timing of omega-3 PUFA administration. In the control LHF group, high-fat feeding for 24 weeks caused a weight gain of ~30 g, which was reduced by 15% in the ω3PL group supplemented with krill oil since the start of the study (Table 1); this reduction in weight gain was mainly due to weight loss during the second half of the study (Figure 2A). Although no changes in average daily food intake (Figure 2B) and cumulative energy intake (Table 1) were observed between the groups, the average feeding efficiency, calculated by dividing the weight gain by the amount of energy consumed each week, was significantly reduced in ω3PL mice (Figure 2C). The adiposity index did not change in response to omega-3 PUFA administration (Table 1), but specifically in the ω3PL group, mesenteric WAT weight was reduced by ~27% (a similar trend was also observed in ω3PL-R mice), while the weight of epididymal WAT was increased in the ω3PL and ω3PL-R groups (Table 1). Despite the increased weight of epididymal WAT in both krill oil-supplemented groups, the average size of adipocytes in this fat depot remained unchanged (Figure 2D) while tissue accumulation of inflammatory macrophages, analyzed by MAC-2/galectin-3 immunodetection (see Supplementary Figure S1) and assessed by CLS counting, was reduced (Figure 2E); no such changes were observed in the ω3TG-R group. With regard to lipid metabolism markers in the circulation, omega-3 PUFAs reduced total cholesterol levels, regardless of the lipid form of their supplementation (Table 1). Given the role of obesity and WAT inflammation in impaired glucose metabolism, we next evaluated the effect of omega-3 PUFAs on glucose homeostasis; it was improved specifically in krill oil-supplemented mice (i.e., ω3PL and ω3PL-R), as evidenced by lower FBG and non-fasting plasma insulin (Table 1). Furthermore, these mice also showed stronger induction of plasma adiponectin levels (Figure 2F) in association with significantly reduced HOMA-IR (Figure 2G).

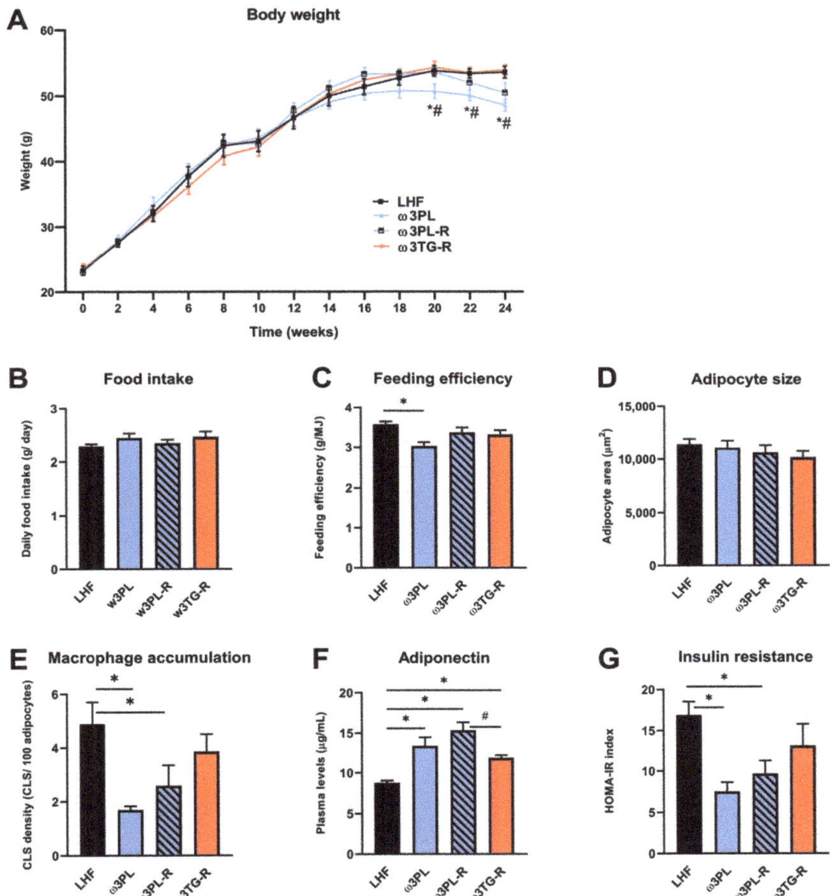

Figure 2. The effect of omega-3 PUFA supplementation on parameters related to energy balance, adipose tissue health and insulin sensitivity: changes in body weight during the study (**A**), average daily food intake (**B**), feeding efficiency (**C**), average size of adipocytes (**D**) and macrophage accumulation in epididymal WAT (**E**), plasma adiponectin levels (**F**), and insulin resistance based on the HOMA-IR index (**G**). Data are means ± SEM (n = 7–8). *, significant effect of omega-3 PUFAs (vs. LHF); #, significant difference from ω3TG-R (One Way ANOVA or Kruskal–Wallis).

Table 1. Energy balance, adiposity, and biochemical parameters in mice fed high-fat diets supplemented or not with omega-3 PUFA concentrates.

	LHF	ω3PL	ω3PL-R	ω3TG-R
Body weight (g)				
Week 0	23.3 ± 0.7	23.3 ± 0.6	23.1 ± 0.5	23.6 ± 0.7
Week 24	53.6 ± 0.9	48.6 ± 0.9 [a]	50.5 ± 1.5	53.9 ± 0.8 [b]
Gain	30.3 ± 0.8	25.3 ± 1.3 [a]	27.4 ± 1.6 [b]	30.2 ± 0.4 [b]
En. intake (MJ/mouse/study)	8.1 ± 0.1	8.1 ± 0.3	7.9 ± 0.3	7.7 ± 0.3
WAT depots (g)				
Epididymal WAT	2.11 ± 0.06	2.54 ± 0.15 [a]	2.52 ± 0.12 [a]	2.07 ± 0.07 [b,c]
Subcutaneous WAT	1.59 ± 0.06	1.45 ± 0.06	1.55 ± 0.06	1.75 ± 0.10
Mesenteric WAT	1.50 ± 0.07	1.09 ± 0.06 [a]	1.36 ± 0.11 [b]	1.48 ± 0.07 [b]
Adiposity index (%)	9.7 ± 0.2	10.5 ± 0.4	10.7 ± 0.4	9.8 ± 0.3
Clinical biochemistry				
TAGs (mmol/L)	1.14 ± 0.12	0.91 ± 0.05	1.02 ± 0.11	0.97 ± 0.04
NEFA (mmol/L)	0.61 ± 0.05	0.57 ± 0.05	0.60 ± 0.07	0.64 ± 0.06
Cholesterol (mmol/L)	6.22 ± 0.18	5.10 ± 0.30 [a]	5.49 ± 0.14 [a]	5.87 ± 0.14 [b]
FBG (mmol/L)	5.19 ± 0.12	4.33 ± 0.09 [a]	4.46 ± 0.21 [a]	5.00 ± 0.20 [b,c]
Insulin (ng/mL)	4.80 ± 0.60	2.65 ± 0.29 [a]	3.92 ± 0.64	5.74 ± 0.66 [b]

Data are means ± SEM (n = 7–8). Except FBG, measured in overnight fasted mice, biochemical parameters were determined in plasma of mice fed ad libitum. [a,b,c] different from LHF, ω3PL, ω3PL-R, respectively (one-way ANOVA or Kruskal–Wallis). FBG, fasting blood glucose; NEFA, non-esterified fatty acids.

Thus, in obese mice kept in a thermoneutral environment, administration of krill oil, unlike omega-3 PUFAs supplemented in the TAG form, led to redistribution of WAT and improvement of its function, which corresponded to positive effects on glucose homeostasis.

3.2. Histological Analysis of NAFLD-Related Phenotypes

We further examined the effect of krill oil and omega-3 PUFAs supplemented as TAGs on the development of NAFLD using biochemical and histological analyses of the liver (Figures 3 and 4).

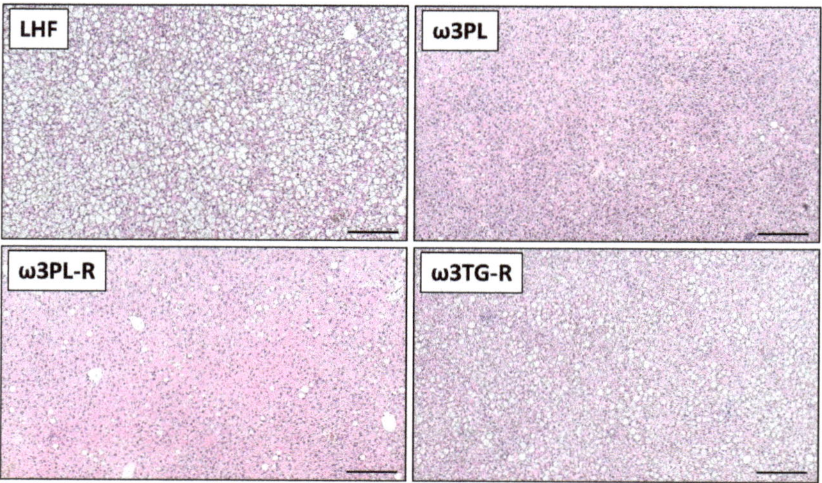

Figure 3. Representative histological sections of liver stained with hematoxylin and eosin. Bars = 200 μm.

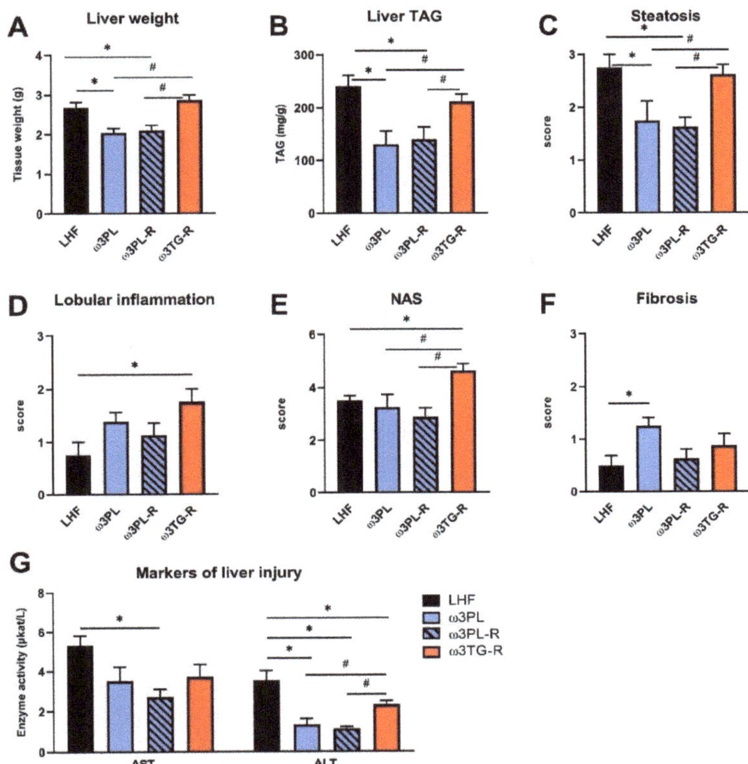

Figure 4. The effect of omega-3 PUFA supplementation on NAFLD-related parameters: liver weight (**A**), liver TAG content (**B**), steatosis (**C**), lobular inflammation (**D**), NAFLD activity score—NAS (**E**), fibrosis (**F**), and plasma AST and ALT levels (**G**). The results presented in panels C–F are based on histological analysis of liver sections. Data are means ± SEM (*n* = 7–8). *, significant effect of omega-3 PUFAs (vs. LHF); #, significant difference from ω3TG-R (one-way ANOVA or Kruskal–Wallis).

Despite having similar plasma TAG levels (Table 1), liver weight (Figure 4A) and the TAG content in the liver (Figure 4B; quantified biochemically) were reduced in both experimental groups that received krill oil compared either to LHF controls or ω3TG-R mice supplemented with omega-3 PUFAs as TAGs. Histological analysis of hematoxylin-eosin-stained liver sections (Figure 3) confirmed reduced levels of steatosis in the livers of ω3PL and ω3PL-R mice compared to the LHF and ω3TG-R groups, where the steatosis score reached almost the maximum value of 3 (Figure 4C). In addition to the degree of steatosis, histological analysis was also used to assess other components of the NAFLD activity score (NAS), which includes lobular inflammation and hepatocyte ballooning. Lobular inflammation was increased in ω3TG-R mice and unchanged in krill oil-supplemented mice compared to LHF-fed controls (Figure 4D), while hepatocyte ballooning was relatively less frequent (score < 0.5) and was similar among the groups (not shown). As a result, the NAS score was higher in ω3TG-R mice and unchanged in ω3PL and ω3PL-R animals compared to LHF-fed controls (Figure 4E). We also evaluated the degree of fibrotic changes in the liver, which was generally relatively low in all groups (Figure 4F); however, ω3PL mice supplemented with krill oil during the development of obesity showed significantly higher fibrosis score compared to control animals fed LHF (Figure 4F). On the other hand, plasma levels of transaminases, especially ALT, were reduced due to krill oil supplementation, with a weaker effect observed in mice receiving omega-3 PUFAs as TAGs (Figure 4G).

These data primarily document the excellent efficacy of krill oil in reducing severe hepatic steatosis induced by administration of a high-fat diet in a thermoneutral environment.

3.3. Analysis of Parameters Related to the Efficacy of Omega-3 PUFAs in the Liver

Next, we analyzed some parameters that may determine the metabolic effects of krill oil in the liver (Figure 5). First, the bioavailability of FAs such as arachidonic acid (ARA), EPA, and DHA, which are substrates for the formation of biologically active lipid mediators, was measured in the neutral (i.e., mainly TAGs) and polar (i.e., mainly PLs) lipid fractions of the liver (Figure 5A,B; for complete data on FAs composition in hepatic TAGs and PLs, see Tables S4 and S5, respectively).

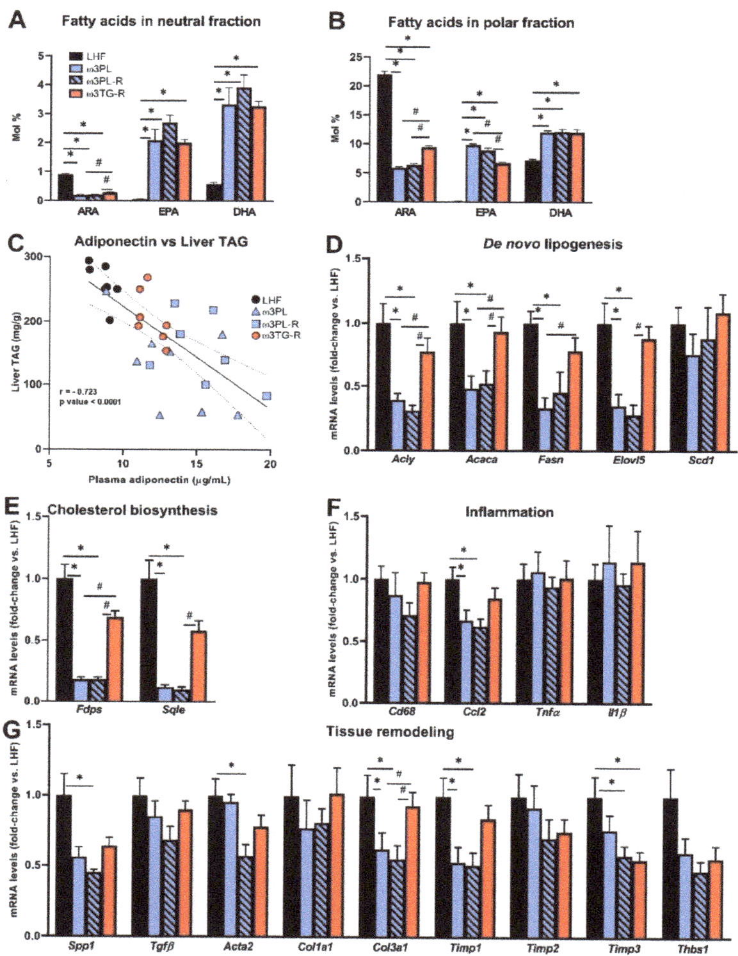

Figure 5. Selected parameters determining the effects of omega-3 PUFAs in the liver: bioavailability of FAs such as arachidonic acid (ARA), eicosapentaenoic acid (EPA) and docosahexaenoic acid (DHA) measured in the neutral (**A**) and polar (**B**) fraction of liver lipids; correlation between plasma adiponectin levels and the degree of TAG accumulation in the liver (**C**); hepatic expression of genes related to DNL (**D**), cholesterol biosynthesis (**E**), inflammation (**F**), and tissue remodeling (**G**). Data are means $\pm$ SEM (n = 7–8). *, significant effect of omega-3 PUFAs (vs. LHF); #, significant difference from ω3TG-R (one-way ANOVA or Kruskal–Wallis).

In general, the relative content of ARA was reduced while the content of EPA and DHA was increased in response to administration of both krill oil and omega-3 PUFAs as TAGs; however, the ARA reduction efficacy in both the neutral and polar fractions was significantly higher in the krill oil-supplemented groups (Figure 5A,B), and the same situation was observed in the case of an increase in EPA primarily in the polar fraction (Figure 5B). We also examined the relationship between plasma adiponectin levels and the degree of TAG accumulation in the liver (Figure 5C), as adiponectin can activate 5′-AMP-activated protein kinase (AMPK) and PPARα and thus stimulate lipid catabolism. Indeed, in the livers of mice fed various LHF-based diets, plasma adiponectin levels and liver TAG levels showed a strong negative correlation ($r = -0.723$; $p < 0.0001$; Figure 5C).

Furthermore, the above findings regarding NAFLD-related phenotypes were then related to changes in gene expression of key enzymes of lipid and cholesterol metabolism, as well as markers of inflammation and tissue remodeling with a known relationship to the development of NAFLD/NASH (Figure 5D–G). Expression of the genes related to DNL (Figure 5D) and cholesterol biosynthesis (Figure 5E) was reduced up to 10-fold in krill oil-supplemented mice (ω3PL and ω3PL-R mice) compared to LHF-fed controls. In contrast, no such changes were observed in ω3TG-R mice receiving omega-3 PUFAs as TAGs (Figure 5D,E). Although histological analysis did not reveal a significant degree of NASH/fibrosis induced in C57BL/6N mice fed the LHF diet in a thermoneutral environment (Figure 4D,F), the expression of both inflammatory genes and tissue remodeling genes in the liver was analyzed in order to see whether supplementation with krill oil or omega-3 PUFAs as TAGs could affect these processes at this level. There were no significant differences between groups in the expression of inflammation-related genes (Figure 5F), with the exception of chemokine (C-C motif) ligand 2 (CCL2; also known as monocyte chemoattractant protein 1), whose expression was reduced in groups ω3PL and ω3PL-R. No consistent effects on the expression of genes related to tissue remodeling were observed (Figure 5G); in general, the administration of krill oil reduced the mRNA levels of some of the measured genes (e.g., Spp1, Col3a1, Timp1, Timp3), with occasional effects (e.g., Timp3) also observed in ω3TG-R mice given omega-3 PUFAs as TAGs. Decreased expression of genes involved in DNL may therefore help explain the beneficial effects of krill oil supplementation on liver fat accumulation, while its effect on the expression of inflammatory genes, and in particular tissue remodeling genes, was not conclusive.

3.4. In Vivo Analyses Related to Liver Function and Insulin Sensitivity

Given the significant reduction in TAG levels in the livers of mice supplemented with krill oil (Figure 4B,C) and a likely reduction in insulin resistance (i.e., HOMA-IR; Figure 2G) in these animals, we initiated an additional experiment (see also Figure 1B), in which we performed a series of in vivo functional assays in mice fed either the Chow, LHF or ω3PL diet for 24 weeks at 30 °C (Figure 6). At the end of the study, mice in the Chow, LHF and ω3PL groups weighed 41.3 ± 1.1, 57.6 ± 0.5, and 54.5 ± 0.8 g ($p < 0.01$ vs. LHF for both other groups), which corresponded to a weight gain of 14.5 ± 1.0, 30.7 ± 0.4, and 27.5 ± 0.6 g ($p < 0.0001$ vs. LHF), respectively. Thus, the weight gain in LHF and ω3PL mice was similar to that observed in the previous experiment (see Table 1). We first evaluated whether lower levels of hepatic steatosis in ω3PL mice, supplemented with krill oil during the development of obesity, can be explained by changes in hepatic production of TAGs contained in VLDL (Figure 6A,B). However, the VLDL-TAG secretion test in overnight fasted mice did not reveal any significant differences in lipemic curves between the groups, especially LHF and ω3PL mice (Figure 6A), although basal plasma TAG levels were decreased by 15% in ω3PL compared to LHF mice (Figure 6B). We further examined whether the potent effect of krill oil supplementation on hepatic steatosis is associated with changes in gluconeogenesis and/or improved insulin sensitivity. Thus, we evaluated the glycemic response to pyruvate injection to determine gluconeogenic activity; glycemic curves (Figure 6C), as well as quantification of the glycemic response based on AUC values (Figure 6D), indicated a reduced level of pyruvate-stimulated gluconeogenesis in both

Chow and ω3PL mice compared to LHF-fed controls. Finally, we used the state-of-the-art hyperinsulinemic-euglycemic clamp technique in combination with a radioactive glucose tracer to analyze whole-body and hepatic insulin sensitivity (Figure 6E,F). Whole-body insulin sensitivity was markedly impaired in obese LHF-fed control animals compared to Chow-fed mice, as documented by changes in glucose infusion rate (GIR; Figure 6F), reflecting the amount of exogenous glucose required to maintain euglycemia during the clamp (i.e., under insulin-stimulated conditions) and which showed a ~2.7-fold reduction in LHF mice.

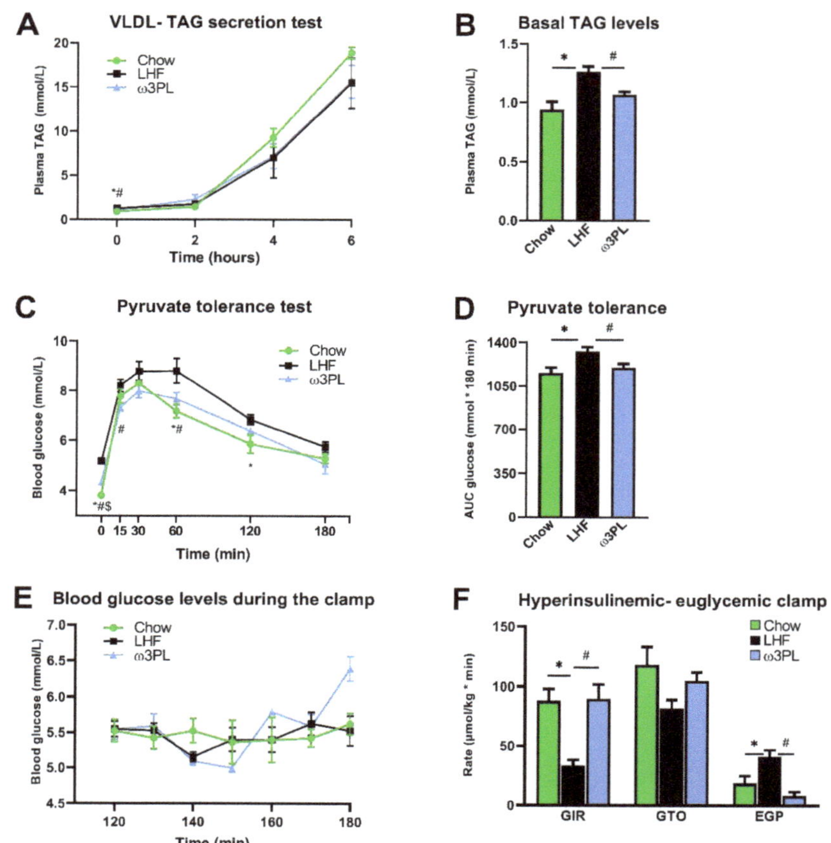

Figure 6. Effect of krill oil supplementation on hepatic VLDL-TAG production (**A**), fasting plasma TAG levels (**B**), glycemia during the pyruvate tolerance test (**C**), the level of pyruvate-driven gluconeogenesis (**D**), as well as glycemia during the last hour of hyperinsulinemic-euglycemic clamp (**E**) and clamp-related parameters including glucose infusion rate (GIR), glucose turnover (GTO) and endogenous glucose production (EGP; **F**). Data are means ± SEM (n = 6–7). *, significant difference between LHF and Chow; #, significant difference between LHF and ω3PL; $, significant difference between ω3PL and Chow (one-way ANOVA or Kruskal–Wallis).

However, the total glucose turnover (GTO) in the organism was not significantly changed in LHF mice compared to the Chow group, primarily because of an ~2.2-fold increase in endogenous glucose production (EGP; Figure 6F). In contrast, krill oil supplementation in ω3PL mice led to normalization of insulin sensitivity at the whole-body level and in the liver, as shown by increased GIR and decreased EGP levels under hy-

DAGs containing SFA (Figure 9B,C), while supplementation of omega-3 PUFAs as TAGs in the ω3TG-R group was ineffective. In contrast, ω3TG-R mice showed increased levels of DAG species containing primarily DHA (e.g., DAG 44:12; DAG 22:6_22:6), with lower increases observed in the krill oil-supplemented groups (Figure 9D). However, DAG species containing at least 1 SFA represented the majority of all DAGs (20 out of 37), while the levels of SFA-containing DAGs showed a very strong correlation with the total levels of all DAGs (Figure 9E). In addition, total levels of short-chain TAGs were analyzed and found to be reduced in both krill oil-supplemented groups (i.e., ω3PL, ω3PL-R) compared to LHF-fed controls (Figure 9F), while the strongest effect was observed at the level of only SFA-containing TAGs (Figure 9G,H). No such effects were observed in ω3TG-R mice supplemented with omega-3 PUFAs as TAGs.

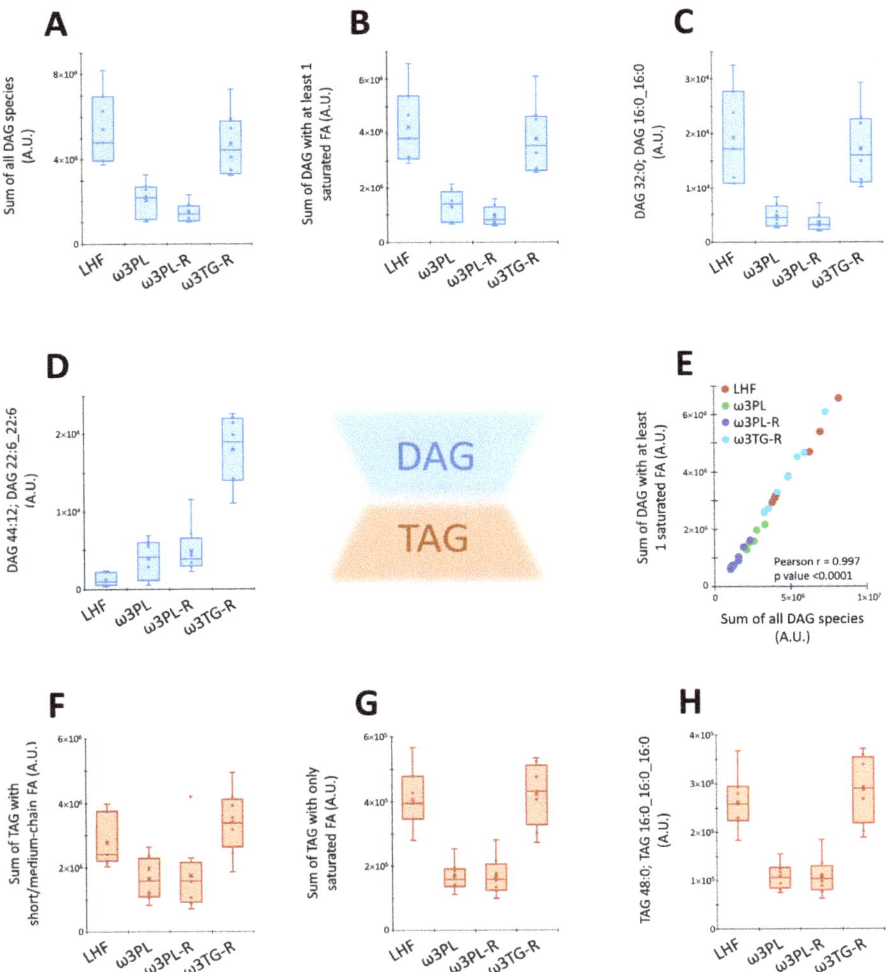

Figure 9. Box plots and correlation analysis for DAGs and TAGs lipid classes in the liver in response to dietary challenges: the sum of all DAG species (**A**; *n* = 37), the sum of DAGs with at least 1 SFA (**B**; *n* = 20), and representative species of DAGs containing either SFA (i.e., DAG 32:0; DAG 16:0_16:0; **C**) or omega-3 PUFAs (i.e., DAG 44:12; DAG 22:6_22:6; **D**). Correlation between the sum of all DAG species (*n* = 37) and the sum of DAGs with at least 1 SFA (*n* = 20; **E**). The sum of TAG species with short/medium-chain FAs (**F**; *n* = 9), the sum of TAG species with only SFAs (**G**; *n* = 3), and a representative SFA-containing TAG species TAG 48:0; TAG 16:0_16:0_16:0 (**H**). Lipid intensities are in arbitrary units (A.U.).

Overall, the above data suggest that krill oil administration has led to profound changes in the levels of complex lipids as well as polar metabolites in the liver, while affecting lipid species that are either involved in the induction of insulin resistance or are established markers of DNL.

4. Discussion

This study aimed to assess the ability of krill oil supplementation to affect NAFLD-related phenotypes in mice with diet-induced obesity and exacerbated NAFLD. At the same time, we wanted to find out whether the effects of krill oil administration on NAFLD are associated with changes in insulin sensitivity, and to look for possible determinants of these effects. Our results clearly demonstrate the ability of krill oil supplementation to alleviate hepatic steatosis, even in a situation when the accumulation of fat in the liver is maximally stimulated due to the combination of high-fat feeding and thermoneutral housing. In contrast, similar doses of omega-3 PUFAs administered via a TAG-based concentrate did not significantly reduce hepatic fat accumulation under the severe obesogenic and steatosis-promoting conditions. In addition, the potent antisteatotic effects of krill oil were observed in a situation when insulin sensitivity in the liver and at the whole-body level was maintained, and which was associated with a hepatic lipidomic signature characterized by reduced concentrations of both short/medium-chain TAGs and total DAGs.

To evaluate the efficacy of krill oil administration in influencing NAFLD-associated phenotypes, we adopted a recently established model of exacerbated NAFLD in C57BL/6J mice [36], which combines the administration of a lard-based high-fat diet with thermoneutral animal housing (i.e., ambient temperature ~30 °C). However, instead of using the "J" substrain of C57BL/6 mice as in Giles et al. [36], we used the "N" substrain (i.e., C57BL/6N mice) due to its apparent ability to accumulate a larger amount of TAGs in the liver when fed a high-fat diet. In fact, when comparing high-fat diet-fed mice of the C57BL/6J [36] and C57BL/6N [46] substrains, kept at 22 °C, C57BL/6N mice accumulated more fat in the liver despite being fed a corn oil-based diet rich in n-6 PUFA, which has a lower potential to induce hepatic steatosis compared to an SFA-rich lard-based diet [9,10,14]. Indeed, in terms of liver fat accumulation, the use of C57BL/6N mice in the current study led to the induction of very severe hepatic steatosis with TAG levels reaching ~250 mg/g, which was much more than in the corresponding group of C57BL/6J mice in the reference study by Giles et al. (i.e., ~130 mg/g in the HFD group at 30 °C; see Figure 2e in [36]). Thus, the use of the C57BL/6N mouse substrain in combination with thermoneutral housing and lard-based high-fat feeding allowed remarkably high accumulation of TAGs in the liver.

Notwithstanding the above differences in liver fat accumulation between the two substrains of C57BL/6 mice, our current study demonstrated the ability of krill oil administered to C57BL/6N mice to induce strong antisteatotic effects even in the presence of pronounced hepatic steatosis. Although some previous studies have already shown the beneficial effects of dietary omega-3 PUFAs as PLs on liver fat accumulation in various rodent models of obesity [24–26,29,30], none of those models reached the level of liver fat accumulation observed in our current study (see above). Even in genetically obese Zucker *fa/fa* rats, the fat content in the liver did not exceed 200 mg/g [25], while in mice fed different high-fat diets [24,26,29,30], it ranged from ~50 to ~160 mg/g, depending on the type and percentage of fat in the diet and the duration of its administration. This work, together with previous studies that involved supplementation of omega-3 PUFAs as PLs using either krill oil [24,25,29] or herring meal extract [26,30], thus demonstrates the excellent ability of krill oil to positively affect liver steatosis, at a wide range of tissue concentrations of TAGs. Importantly, in our present study, the antisteatotic effects of krill oil supplementation were observed regardless of whether krill oil was added to the LHF diet from the very beginning of the dietary interventions or after previous administration of the LHF diet when the animals were already obese.

Krill oil, containing significant amounts of PLs, especially phosphatidylcholine [31], was used in the current study to administer primarily EPA and DHA. The type of krill oil used in the current study is characterized by the presence of two main fractions, i.e., phosphatidylcholines and TAGs, which represent 49 and 28%; furthermore, free FAs, DAGs, ether-linked phosphatidylcholines, cholesterol, phosphatidylethanolamines, phosphatidylinositols and lysophosphatidylcholines represent minor fractions of 4.9%, 3.5%, 3.5%, 3.4%, 3.2%, 2.3%, and 2.0%, respectively [31]. Because TAGs, along with ethyl esters, represent lipid classes traditionally used for supplying omega-3 PUFAs into the organism and to treat NAFLD in humans [21,23], we also included in our current study a group of obese mice given EPA and DHA as a TAG-based concentrate. Specifically, we used Epax 3000 TG concentrate with an EPA and DHA content of 29%, which corresponds to the relative content of these FAs in krill oil [47]. Thus, similar amounts of lard in the LHF diet had to be replaced by one or the other concentrate in order to prepare respective supplemented diets. Furthermore, the EPA:DHA ratio was approximately 1.6:1 in both krill oil and Epax 3000 TG, which facilitates the interpretation of our results compared to previous reports (e.g., [26,29]), where EPA and DHA content, as well as their ratio, differed significantly between the omega-3 PUFA concentrates based on TAGs or PLs. However, despite the similarity in the concentration and ratio of EPA and DHA, krill oil and the TAG-based omega-3 PUFA concentrate differed dramatically in terms of their effects on hepatic steatosis. As determined by biochemical analysis of TAG content in the tissue and histological evaluation (i.e., steatosis score), dietary supplementation with krill oil resulted in a 42% reduction in TAG (glycerolipids) accumulation in the liver of obese mice (i.e., ω3PL-R mice), while in mice with omega-3 PUFAs supplemented as TAGs (i.e., ω3TG-R mice) only an insignificant decrease of 5–12% was found. While this difference cannot be explained by the effects on energy intake, body weight or overall adiposity, adipose tissue functionality was improved specifically in krill oil-supplemented mice. A more pronounced reduction in macrophage accumulation in epididymal WAT of ω3PL-R mice was accompanied by higher plasma adiponectin levels (up to ~15 μg/mL) compared to their counterparts treated with the same dose of omega-3 PUFAs given as TAGs. These data are consistent with previous reports documenting the excellent efficacy of krill oil (vs. omega-3 PUFAs as TAGs; [29]) and its dose-dependent effects [24,29] in stimulating plasma adiponectin levels in high-fat diet-fed mice housed under standard thermal conditions. In this context, krill oil administered at approximately 3-fold lower dose to mice fed a 21% fat diet caused an increase in adiponectin levels to only 7.5 μg/mL [24], which is about half of the values achieved in our current study (see above). The role of elevated adiponectin levels in the antisteatotic effects of krill oil supplementation in our study is supported by the presence of a strong negative correlation between plasma adiponectin levels and the degree of liver TAG accumulation. A similar relationship has been shown, for example, in type 2 diabetic patients before and after treatment with insulin sensitizers thiazolidinediones, which resulted in a reduction in liver fat and improved insulin sensitivity while plasma adiponectin increased [48]. Thus, improved WAT function combined with a substantial increase in plasma adiponectin levels may play a role in the potent effects of krill oil supplementation on both liver steatosis and insulin sensitivity in our mice with exacerbated NAFLD.

Adiponectin levels are negatively associated with hepatic and peripheral insulin resistance and hepatic fat content [49,50]. It is known that adiponectin activates AMPK in both skeletal muscle and liver [51]. Activation of AMPK in turn leads to inhibition of acetyl-CoA carboxylase, a key enzyme in the DNL pathway, as well as to induction of FA oxidation and suppression of lipogenic enzymes [51–53]. This is in line with our current data and results from previous rodent studies using various forms of omega-3 PUFAs as PLs [29,30,33,34], which show decreased gene expression primarily within the DNL and cholesterol biosynthesis pathways. In addition to effects on FA metabolism pathways, AMPK is also important for the suppressive effect of adiponectin on hepatic glucose production and for maintaining normal fasting glucose levels [54], as well as for the beneficial effect of omega-3 PUFAs on hepatic insulin sensitivity [53]. Our results showing increased plasma adiponectin associated with decreased TAG accumulation and lipogenic gene expression in the liver are therefore consistent with the involvement of the adiponectin-

AMPK axis in the antisteatotic effects of krill oil supplementation. Furthermore, activation of this axis could also explain the observed reduction in tissue DAG levels (especially those containing SFAs) in krill oil-supplemented mice. This effect may be directly related to the improvement of hepatic insulin sensitivity, as DAGs are known to be strongly involved in the development of hepatic insulin resistance [53,55]. Because SFA-containing DAGs represented the majority of DAG species in the liver, a marked reduction in total DAG levels in mice supplemented with krill oil may be a direct consequence of its inhibitory effects on DNL (see below).

In addition to elevated adiponectin levels, there are likely to be other mechanisms by which krill oil effectively alleviates exacerbated liver steatosis. Although krill oil supplementation does not seem to change the level of VLDL-TAG secretion (our current data and [34]), it can reduce the mitochondrial citrate carrier activity, as previously observed in rats fed a lard-based high-fat diet [35], and which also showed reduced activities of DNL enzymes such as acetyl-CoA carboxylase and fatty acid synthase. This carrier acts upstream of cytosolic lipogenic processes [56], and its inhibition could thus explain the strong antilipogenic properties of krill oil. In fact, our lipidomics data from the livers of mice fed with krill oil document significantly reduced levels of short/medium-chain TAGs, a subset of TAGs previously proposed as a DNL marker [57]. Therefore, inhibition of this pathway may be one of the main mechanisms of the antisteatotic effects of krill oil. Furthermore, krill oil administration can lead to effective stimulation of FA oxidation in the liver [35], which in turn may be related to the ability of EPA, but not DHA, to increase FA oxidation while inhibiting 1,2-diacylglycerol esterification and thus TAG synthesis in hepatocytes [58]. This is in line with our current and previous [29] results documenting the improved bioavailability of EPA at the level of liver PLs. Furthermore, a recent study on mice fed a high-fat diet based on corn oil suggests that intestinal FA oxidation, which was more effectively stimulated by krill oil compared to omega-3 PUFAs supplemented as TAGs, could also be involved in the antisteatotic effects of this marine oil [59]. It is worth mentioning that other bioactive constituents are present in krill oil, which may possess antisteatotic and insulin-sensitizing properties. Our metabolomics analysis revealed that stachydrine and trigonelline are among the top two polar metabolites that most distinguish the krill oil-supplemented groups from LHF-fed controls, as well as mice supplemented with omega-3 PUFAs as TAGs. Here we show that both of these alkaloids are enriched in the diet supplemented with krill oil, and both have previously been shown to have positive effects on NAFLD, probably by restoring hepatic autophagy [60,61]. The increase in hepatic concentrations of TMAO in mice fed krill oil was probably due to its increased biosynthesis from choline by intestinal bacteria [62]. Moreover, palmitoleic acid contained in krill oil may contribute not only to the positive effects of this marine oil on glucose homeostasis and insulin sensitivity [29], but also on liver steatosis due to its stimulatory effects on PPARα and AMPK activation [63]. The complex composition of krill oil and the role of its constituents in influencing liver fat accumulation and insulin sensitivity is shown in Figure 10 below. Recently, 3-carboxy-4-methyl-5-propyl-2-furanpropanoic acid (CMPF) has been reported as a plasma metabolite whose levels increased with omega-3 PUFAs intake and which was able to alleviate hepatic steatosis when administered to mice [64,65]. Using our metabolomics method we detected CMPF at very low signal intensities in the liver samples. Higher fold changes of 1.7 and 1.9 were observed in the ω3TG-R group compared to the ω3PL and ω3PL-R groups, respectively. However, in the control LHF group, high biological variability of CMPF was noticed, and thus, this metabolite was not ranked among the most discriminating ones.

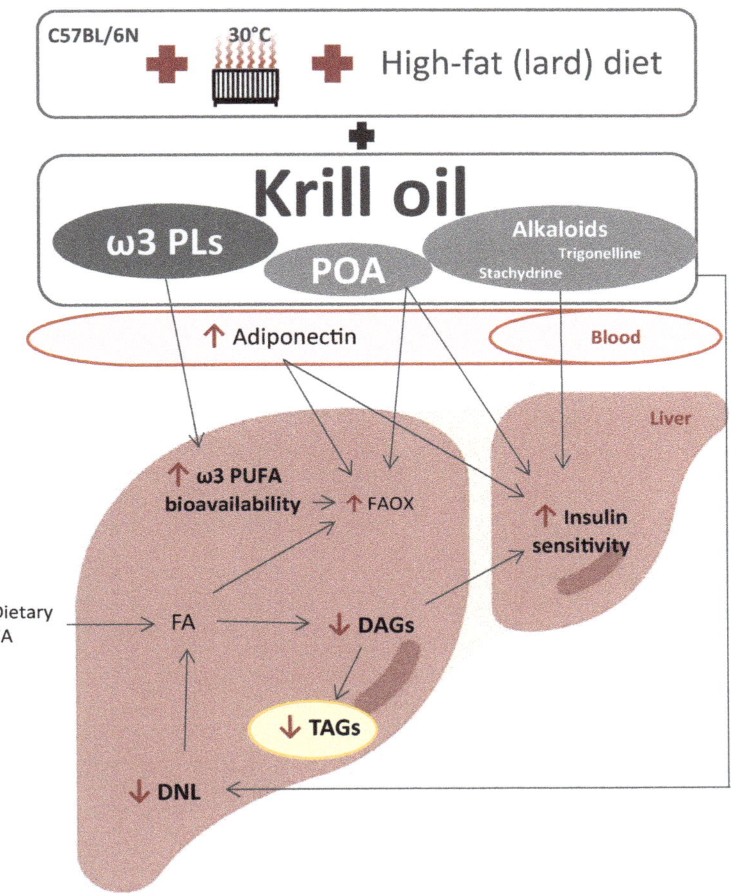

Figure 10. The potential mechanisms involved in the effects of krill oil supplementation on liver fat accumulation and insulin sensitivity in a mouse model of exacerbated hepatic steatosis induced in C57BL/6N mice fed a high-fat (lard) diet in a thermoneutral environment. The effects of krill oil are determined not only by omega-3 PUFA-containing PLs (ω3 PLs) in this marine oil, but also by its other bioactive constituents, including palmitoleic acid (POA) and the alkaloids stachydrine and trigonelline, and may involve direct or indirect mechanisms. The livers of mice fed a high-fat diet supplemented with krill oil have markedly reduced TAG accumulation and improved insulin sensitivity, which is associated with increased bioavailability of omega-3 PUFAs, suppressed DNL, decreased tissue DAG levels, and stimulated FA oxidation (FAOX). While many of these changes may be due to indirect mechanisms based on the beneficial effect of krill oil on WAT functionality associated with markedly elevated plasma adiponectin levels, direct mechanisms may include the effect of stachydrine and/or trigonelline, i.e., alkaloids contained in krill oil, which have previously been shown to have positive effects on NAFLD, presumably by restoring hepatic autophagy.

It is not clear why supplementation with omega-3 PUFAs in the form of a TAG-based concentrate did not reduce liver fat in this model of exacerbated hepatic steatosis, despite the fact that the content of both EPA and DHA in the liver was significantly increased. In this regard, we can speculate that the stronger effects of krill oil on liver steatosis are based on a combination of a number of factors, including higher adiponectin levels along with better bioavailability of EPA in liver tissue, as well as the specific effect of alkaloids contained in krill oil. In addition, choline contained as phosphatidylcholine in krill oil can also contribute to the strong antisteatotic effects of this oil in the liver, as compared to omega-3 PUFAs supplemented as TAGs. Poor availability of hepatic choline/phosphatidylcholine is known to promote steatosis by various mechanisms, including increased DNL and impaired synthesis and secretion of hepatic VLDL (reviewed in [66]). In this context, our previous study in mice fed a corn oil-based high-fat diet showed that the antisteatotic effects of dietary phosphatidylcholine in the liver were unique to PLs containing DHA and EPA, whereas these effects were not present in animals fed soy-derived phosphatidylcholine, which contained mainly PUFAs of n-6 series such as linoleic acid [30].

Among the main weaknesses of our study is the fact that, although we used an established model of exacerbated NAFLD, whose characteristics should include NASH and liver fibrosis [36], we were not able to sufficiently induce these characteristics in our experimental mice. While the exact cause is not obvious, it may be related to the fact that a different substrain of C57BL/6 mice was used in our current study. Thus, despite a maximum steatosis score of about 3, LHF-fed control mice of the C57BL/6N substrain showed only minimal lobular inflammation (score < 1), and the overall NAS score of less than 4. This is in sharp contrast to the reference study performed on C57BL/6J mice, where the mean NAS score was ~7 [36]. Therefore, it was not possible to assess the effect of various forms of omega-3 PUFA supplementation on NASH/fibrosis in our current study. Interestingly, however, in LHF-fed control mice, plasma levels of ALT, a marker of liver damage, were almost comparable in our and the reference study. Therefore, the marked decrease in plasma ALT levels in the krill oil-supplemented groups may be due to both the potential protective effect of this oil on liver tissue and its inhibitory effect on hepatic gluconeogenesis [67]. Furthermore, the strengths of our study include: (i) mouse model with marked hepatic steatosis; (ii) evaluation of the relative efficacy of krill oil versus omega-3 TAGs in terms of effects on liver fat; (iii) comprehensive methodological approach, including various in vivo techniques such as hyperinsulinemic-euglycemic clamps, which revealed a number of potential mechanisms of action of krill oil on liver steatosis; and (iv) metabolomic analysis that identified, in addition to omega-3 PLs, other constituents of krill oil that may contribute to the potent antisteatotic effects of this oil.

5. Conclusions

By using C57BL/6N mice in combination with thermoneutral housing and lard-based high-fat feeding, we achieved remarkably high levels of TAG accumulation in the liver. Despite these extreme conditions, severe hepatic steatosis was markedly reduced in response to krill oil administration, but not in response to omega-3 PUFAs using a TAG-based concentrate. The potent antisteatotic effects of krill oil, which have been observed in both the prevention and reversal of hepatic steatosis, were associated with improved insulin sensitivity in the liver and at the systemic level. Mechanistically, high plasma adiponectin levels, as well as improved EPA bioavailability, strong repression of DNL, and decreased levels of DAGs in the liver may explain the above beneficial effects of krill oil on liver fat and insulin sensitivity. Furthermore, the role of polar metabolites contained in krill oil, including alkaloids trigonelline and stachydrine, cannot be excluded. Thus, our results suggest that in addition to omega-3 PUFAs contained in PLs, other constituents of krill oil may contribute to its strong antisteatotic effects in the liver.

Supplementary Materials: The following are available online at https://www.mdpi.com/2072-6643/13/2/437/s1, Table S1: Macronutrient composition of the experimental diets; Table S2: Composition of fatty acids in dietary lipids; Table S3: Gene names and sequences of the oligonucleotide primers; Table S4: Fatty acid composition of the neutral lipid fraction in the liver; Table S5: Fatty acid composition of the polar lipid fraction in the liver; Table S6: Concentration of trigonelline and stachydrine in experimental diets; Table S7: List of annotated complex lipids and polar metabolites in liver samples; Figure S1: Representative histological sections of epididymal white adipose tissue from mice housed in a thermoneutral environment and fed various experimental diets for 24 weeks.

Author Contributions: Conceptualization, J.K. and M.R.; Methodology, T.C., K.B., P.Z., K.L., O.K. and M.R.; Software, T.C.; Validation, T.C., P.Z. and M.R.; Formal analysis, G.S., V.K., T.C., I.I., P.Z., O.H., K.L., O.K., and M.R.; Investigation, G.S., V.K., T.C., I.I., K.B., M.O., P.Z., P.K., O.H., K.L., and M.R.; Resources, M.R.; Data curation, G.S. and M.R.; Writing—original draft preparation, G.S. and M.R.; Writing—review and editing, G.S., V.K., K.B., O.H., K.L., A.G., O.K., J.K, and M.R.; Visualization, G.S., T.C., and M.R.; Supervision, M.R.; Project administration, M.R.; Funding acquisition, J.K. and M.R. All authors have read and agreed to the published version of the manuscript.

Funding: This research was funded by the Czech Science Foundation (grant no. 17-11027S), and the project FOIE GRAS, which has received funding from the European Union's Horizon 2020 Research and Innovation framework, under the Marie Skłodowska-Curie Grant Agreement (no. 722619; recipient G.S.).

Institutional Review Board Statement: The experiments were performed according to the European Guidelines for the care and use of laboratory animals (Directive 2010/63/EU), and approved by the Animal Care and Use Committee of the Institute of Physiology CAS (Approval no. 81/2016).

Informed Consent Statement: Not applicable.

Data Availability Statement: Not applicable.

Acknowledgments: We would like to thank Adela Krejcarkova for technical assistance with hyperinsulinemic-euglycemic clamps. We acknowledge the provision of Epax 3000 TG product and krill oil by Epax Norway AS (Ålesund, Norway) and Rimfrost AS (Ålesund, Norway), respectively.

Conflicts of Interest: The authors declare no conflict of interest.

References

1. Tiniakos, D.G.; Vos, M.B.; Brunt, E.M. Nonalcoholic fatty liver disease: Pathology and pathogenesis. *Annu. Rev. Pathol.* **2010**, *5*, 145–171. [CrossRef] [PubMed]
2. Fabbrini, E.; Sullivan, S.; Klein, S. Obesity and nonalcoholic fatty liver disease: Biochemical, metabolic, and clinical implications. *Hepatology* **2010**, *51*, 679–689. [CrossRef] [PubMed]
3. Younossi, Z.; Tacke, F.; Arrese, M.; Chander Sharma, B.; Mostafa, I.; Bugianesi, E.; Wai-Sun Wong, V.; Yilmaz, Y.; George, J.; Fan, J.; et al. Global perspectives on nonalcoholic fatty liver disease and nonalcoholic steatohepatitis. *Hepatology* **2019**, *69*, 2672–2682. [CrossRef] [PubMed]
4. Marchesini, G.; Bugianesi, E.; Forlani, G.; Cerrelli, F.; Lenzi, M.; Manini, R.; Natale, S.; Vanni, E.; Villanova, N.; Melchionda, N.; et al. Nonalcoholic fatty liver, steatohepatitis, and the metabolic syndrome. *Hepatology* **2003**, *37*, 917–923. [CrossRef]
5. Gastaldelli, A.; Cusi, K. From NASH to diabetes and from diabetes to NASH: Mechanisms and treatment options. *JHEP Rep.* **2019**, *1*, 312–328. [CrossRef]
6. Younossi, Z.; Tampi, R.; Racilla, A.; Qiu, Y.; Burns, L.; Younossi, I.; Nader, F. Economic and clinical burden of non-alcoholic steatohepatitis in patients with type II diabetes in the United States. *Diabetes Care* **2020**, *43*, 283–289. [CrossRef]
7. EASL; Marchesini, G.; Day, C.P.; Dufour, J.-F.; Canbay, A.; Nobili, V.; Ratziu, V.; Tilg, H.; EASD; Roden, M.; et al. EASL–EASD–EASO Clinical Practice Guidelines for the management of non-alcoholic fatty liver disease. *J. Hepatol.* **2016**, *64*, 1388–1402. [CrossRef]
8. Bray, G.A.; Krauss, R.M. Overfeeding of polyunsaturated versus saturated fatty acids reduces ectopic fat. *Diabetes* **2014**, *63*, 2222–2224. [CrossRef]
9. Rosqvist, F.; Iggman, D.; Kullberg, J.; Cedernaes, J.; Johansson, H.-E.; Larsson, A.; Johansson, L.; Ahlström, H.; Arner, P.; Dahlman, I.; et al. Overfeeding polyunsaturated and saturated fat causes distinct effects on liver and visceral fat accumulation in humans. *Diabetes* **2014**, *63*, 2356–2368. [CrossRef]
10. Bjermo, H.; Iggman, D.; Kullberg, J.; Dahlman, I.; Johansson, L.; Persson, L.; Berglund, J.; Pulkki, K.; Basu, S.; Uusitupa, M.; et al. Effects of n-6 PUFAs compared with SFAs on liver fat, lipoproteins, and inflammation in abdominal obesity: A randomized controlled trial. *Am. J. Clin. Nutr.* **2012**, *95*, 1003–1012. [CrossRef]
11. Luukkonen, P.K.; Sädevirta, S.; Zhou, Y.; Kayser, B.; Ali, A.; Ahonen, L.; Lallukka, S.; Pelloux, V.; Gaggini, M.; Jian, C.; et al. Saturated fat is more metabolically harmful for the human liver than unsaturated fat or simple sugars. *Diabetes Care* **2018**, *41*, 1732–1739. [CrossRef] [PubMed]
12. Roche, H.M.; Gibney, M.J. Effect of long-chain n−3 polyunsaturated fatty acids on fasting and postprandial triacylglycerol metabolism. *Am. J. Clin. Nutr.* **2000**, *71*, 232s–237s. [CrossRef] [PubMed]
13. Schuchardt, J.P.; Neubronner, J.; Kressel, G.; Merkel, M.; Von Schacky, C.; Hahn, A. Moderate doses of EPA and DHA from re-esterified triacylglycerols but not from ethyl-esters lower fasting serum triacylglycerols in statin-treated dyslipidemic subjects: Results from a six month randomized controlled trial. *Prostaglandins Leukot. Essent. Fat. Acids* **2011**, *85*, 381–386. [CrossRef] [PubMed]

14. Pavlisova, J.; Bardova, K.; Stankova, B.; Tvrzicka, E.; Kopecky, J.; Rossmeisl, M. Corn oil versus lard: Metabolic effects of omega-3 fatty acids in mice fed obesogenic diets with different fatty acid composition. *Biochimie* **2016**, *124*, 150–162. [CrossRef] [PubMed]
15. Flachs, P.; Rossmeisl, M.; Bryhn, M.; Kopecký, J. Cellular and molecular effects of n−3 polyunsaturated fatty acids on adipose tissue biology and metabolism. *Clin. Sci.* **2009**, *116*, 1–16. [CrossRef]
16. Glass, C.K.; Olefsky, J.M. Inflammation and lipid signaling in the etiology of insulin resistance. *Cell Metab.* **2012**, *15*, 635–645. [CrossRef]
17. Calder, P.C. Marine omega-3 fatty acids and inflammatory processes: Effects, mechanisms and clinical relevance. *Biochim. Biophys. Acta Mol. Cell Biol. Lipids* **2015**, *1851*, 469–484. [CrossRef]
18. Green, C.J.; Pramfalk, C.; Charlton, C.A.; Gunn, P.J.; Cornfield, T.; Pavlides, M.; Karpe, F.; Hodson, L. Hepatic de novo lipogenesis is suppressed and fat oxidation is increased by omega-3 fatty acids at the expense of glucose metabolism. *BMJ Open Diabetes Res. Care* **2020**, *8*, e000871. [CrossRef]
19. Sanderson, L.M.; De Groot, P.J.; Hooiveld, G.J.; Koppen, A.; Kalkhoven, E.; Müller, M.; Kersten, S. Effect of synthetic dietary triglycerides: A novel research paradigm for nutrigenomics. *PLoS ONE* **2008**, *3*, e1681. [CrossRef]
20. Parker, H.M.; Johnson, N.A.; Burdon, C.A.; Cohn, J.S.; O'Connor, H.T.; George, J. Omega-3 supplementation and non-alcoholic fatty liver disease: A systematic review and meta-analysis. *J. Hepatol.* **2012**, *56*, 944–951. [CrossRef]
21. Scorletti, E.; Bhatia, L.; McCormick, K.G.; Clough, G.F.; Nash, K.; Hodson, L.; Moyses, H.E.; Calder, P.C.; Byrne, C.D.; Study, W. Effects of purified eicosapentaenoic and docosahexaenoic acids in nonalcoholic fatty liver disease: Results from the Welcome* study. *Hepatology* **2014**, *60*, 1211–1221. [CrossRef] [PubMed]
22. Argo, C.K.; Patrie, J.T.; Lackner, C.; Henry, T.D.; De Lange, E.E.; Weltman, A.L.; Shah, N.L.; Al-Osaimi, A.M.; Pramoonjago, P.; Jayakumar, S.; et al. Effects of n-3 fish oil on metabolic and histological parameters in NASH: A double-blind, randomized, placebo-controlled trial. *J. Hepatol.* **2015**, *62*, 190–197. [CrossRef] [PubMed]
23. De Castro, G.S.; Calder, P.C. Non-alcoholic fatty liver disease and its treatment with n-3 polyunsaturated fatty acids. *Clin. Nutr.* **2018**, *37*, 37–55. [CrossRef] [PubMed]
24. Tandy, S.; Chung, R.W.S.; Wat, E.; Kamili, A.; Berge, K.; Griinari, M.; Cohn, J.S. Dietary krill oil supplementation reduces hepatic steatosis, glycemia, and hypercholesterolemia in high-fat-fed mice. *J. Agric. Food Chem.* **2009**, *57*, 9339–9345. [CrossRef]
25. Batetta, B.; Griinari, M.; Carta, G.; Murru, E.; Ligresti, A.; Cordeddu, L.; Giordano, E.; Sanna, F.; Bisogno, T.; Uda, S.; et al. endocannabinoids may mediate the ability of (n-3) fatty acids to reduce ectopic fat and inflammatory mediators in obese zucker rats. *J. Nutr.* **2009**, *139*, 1495–1501. [CrossRef]
26. Rossmeisl, M.; Jilkova, Z.M.; Kuda, O.; Jelenik, T.; Medrikova, D.; Stankova, B.; Kristinsson, B.; Haraldsson, G.G.; Svensen, H.; Stoknes, I.; et al. Metabolic effects of n-3 PUFA as phospholipids are superior to triglycerides in mice fed a high-fat diet: Possible role of endocannabinoids. *PLoS ONE* **2012**, *7*, e38834. [CrossRef]
27. Ibrahim, S.H.; Hirsova, P.; Malhi, H.; Gores, G.J. Animal models of nonalcoholic steatohepatitis: Eat, delete, and inflame. *Dig. Dis. Sci.* **2016**, *61*, 1325–1336. [CrossRef]
28. Le Grandois, J.; Marchioni, E.; Zhao, M.; Giuffrida, F.; Ennahar, S.; Bindler, F. Investigation of natural phosphatidylcholine sources: Separation and identification by liquid chromatography−electrospray ionization−tandem mass spectrometry (LC−ESI−MS2) of molecular species. *J. Agric. Food Chem.* **2009**, *57*, 6014–6020. [CrossRef]
29. Rossmeisl, M.; Pavlisova, J.; Bardova, K.; Kalendova, V.; Buresova, J.; Kuda, O.; Kroupova, P.; Stankova, B.; Tvrzicka, E.; Fiserova, E.; et al. Increased plasma levels of palmitoleic acid may contribute to beneficial effects of Krill oil on glucose homeostasis in dietary obese mice. *Biochim. Biophys. Acta Mol. Cell Biol. Lipids* **2020**, *1865*, 158732. [CrossRef]
30. Rossmeisl, M.; Medrikova, D.; Van Schothorst, E.M.; Pavlisova, J.; Kuda, O.; Hensler, M.; Bardova, K.; Flachs, P.; Stankova, B.; Vecka, M.; et al. Omega-3 phospholipids from fish suppress hepatic steatosis by integrated inhibition of biosynthetic pathways in dietary obese mice. *Biochim. Biophys. Acta Mol. Cell Biol. Lipids* **2014**, *1841*, 267–278. [CrossRef]
31. Paluchova, V.; Vik, A.; Cajka, T.; Brezinova, M.; Brejchova, K.; Bugajev, V.; Draberova, L.; Draber, P.; Buresova, J.; Kroupova, P.; et al. Triacylglycerol-rich oils of marine origin are optimal nutrients for induction of polyunsaturated docosahexaenoic acid ester of hydroxy linoleic acid (13-DHAHLA) with anti-inflammatory properties in mice. *Mol. Nutr. Food Res.* **2020**, *64*, e1901238. [CrossRef] [PubMed]
32. Schuchardt, J.P.; Hahn, A. Bioavailability of long-chain omega-3 fatty acids. *Prostaglandins Leukot. Essent. Fat. Acids* **2013**, *89*, 1–8. [CrossRef] [PubMed]
33. Burri, L.; Berge, K.; Wibrand, K.; Berge, R.K.; Barger, J.L. Differential effects of krill oil and fish oil on the hepatic transcriptome in mice. *Front. Genet.* **2011**, *2*, 45. [CrossRef] [PubMed]
34. Tillander, V.; Bjørndal, B.; Burri, L.; Bohov, P.; Skorve, J.; Berge, R.K.; Alexson, S.E.H. Fish oil and krill oil supplementations differentially regulate lipid catabolic and synthetic pathways in mice. *Nutr. Metab.* **2014**, *11*, 20. [CrossRef] [PubMed]
35. Ferramosca, A.; Conte, A.; Burri, L.; Berge, K.; De Nuccio, F.; Giudetti, A.M.; Zara, V. A krill oil supplemented diet suppresses hepatic steatosis in high-fat fed rats. *PLoS ONE* **2012**, *7*, e38797. [CrossRef]
36. Giles, D.A.; Moreno-Fernandez, M.E.; Stankiewicz, T.E.; Graspeuntner, S.; Cappelletti, M.; Wu, D.; Mukherjee, R.; Chan, C.C.; Lawson, M.J.; Klarquist, J.; et al. Thermoneutral housing exacerbates nonalcoholic fatty liver disease in mice and allows for sex-independent disease modeling. *Nat. Med.* **2017**, *23*, 829–838. [CrossRef]

37. Kleiner, D.E.; Brunt, E.M.; Van Natta, M.; Behling, C.; Contos, M.J.; Cummings, O.W.; Ferrell, L.D.; Liu, Y.-C.; Torbenson, M.S.; Unalp-Arida, A.; et al. Design and validation of a histological scoring system for nonalcoholic fatty liver disease. *Hepatology* **2005**, *41*, 1313–1321. [CrossRef]

38. Cinti, S.; Mitchell, G.; Barbatelli, G.; Murano, I.; Ceresi, E.; Faloia, E.; Wang, S.; Fortier, M.; Greenberg, A.S.; Obin, M.S. Adipocyte death defines macrophage localization and function in adipose tissue of obese mice and humans. *J. Lipid Res.* **2005**, *46*, 2347–2355. [CrossRef]

39. Obrowsky, S.; Chandak, P.G.; Patankar, J.V.; Povoden, S.; Schlager, S.; Kershaw, E.E.; Bogner-Strauss, J.G.; Hoefler, G.; Levak-Frank, S.; Kratky, D. Adipose triglyceride lipase is a TG hydrolase of the small intestine and regulates intestinal PPARα signaling. *J. Lipid Res.* **2013**, *54*, 425–435. [CrossRef]

40. Flachs, P.; Rühl, R.; Hensler, M.; Janovska, P.; Zouhar, P.; Kus, V.; Jilkova, Z.M.; Papp, E.; Kuda, O.; Svobodova, M.; et al. Synergistic induction of lipid catabolism and anti-inflammatory lipids in white fat of dietary obese mice in response to calorie restriction and n-3 fatty acids. *Diabetologia* **2011**, *54*, 2626–2638. [CrossRef]

41. Oseeva, M.; Paluchova, V.; Zacek, P.; Janovska, P.; Mráček, T.; Rossmeisl, M.; Hamplova, D.; Cadova, N.; Stohanzlova, I.; Flachs, P.; et al. Omega-3 index in the Czech Republic: No difference between urban and rural populations. *Chem. Phys. Lipids* **2019**, *220*, 23–27. [CrossRef] [PubMed]

42. Paluchova, V.; Oseeva, M.; Brezinova, M.; Cajka, T.; Bardova, K.; Adamcova, K.; Zacek, P.; Brejchova, K.; Balas, L.; Chodounska, H.; et al. Lipokine 5-PAHSA is regulated by adipose triglyceride lipase and primes adipocytes for de novo lipogenesis in mice. *Diabetes* **2020**, *69*, 300–312. [CrossRef] [PubMed]

43. Brezinova, M.; Cajka, T.; Oseeva, M.; Stepan, M.; Dadova, K.; Rossmeislova, L.; Matous, M.; Siklova, M.; Rossmeisl, M.; Kuda, O. Exercise training induces insulin-sensitizing PAHSAs in adipose tissue of elderly women. *Biochim. Biophys. Acta Mol. Cell Biol. Lipids* **2020**, *1865*, 158576. [CrossRef] [PubMed]

44. Chong, J.; Soufan, O.; Li, C.; Caraus, I.; Li, S.; Bourque, G.; Wishart, D.S.; Xia, J. MetaboAnalyst 4.0: Towards more transparent and integrative metabolomics analysis. *Nucleic Acids Res.* **2018**, *46*, W486–W494. [CrossRef] [PubMed]

45. Faul, F.; Erdfelder, E.; Lang, A.-G.; Buchner, A. G*Power 3: A flexible statistical power analysis program for the social, behavioral, and biomedical sciences. *Behav. Res. Methods* **2007**, *39*, 175–191. [CrossRef]

46. Kuda, O.; Jelenik, T.; Jilkova, Z.; Flachs, P.; Rossmeisl, M.; Hensler, M.; Kazdova, L.; Ogston, N.; Baranowski, M.; Gorski, J.; et al. n-3 Fatty acids and rosiglitazone improve insulin sensitivity through additive stimulatory effects on muscle glycogen synthesis in mice fed a high-fat diet. *Diabetologia* **2009**, *52*, 941–951. [CrossRef]

47. Tou, J.C.; Jaczynski, J.; Chen, Y.-C. Krill for human consumption: Nutritional value and potential health benefits. *Nutr. Rev.* **2007**, *65*, 63–77. [CrossRef]

48. Bajaj, M.; Suraamornkul, S.; Piper, P.; Hardies, L.J.; Glass, L.; Cersosimo, E.; Pratipanawatr, T.; Miyazaki, Y.; DeFronzo, R.A. Decreased plasma adiponectin concentrations are closely related to hepatic fat content and hepatic insulin resistance in pioglitazone-treated type 2 diabetic patients. *J. Clin. Endocrinol. Metab.* **2004**, *89*, 200–206. [CrossRef]

49. Bugianesi, E.; Pagotto, U.; Manini, R.; Vanni, E.; Gastaldelli, A.; De Iasio, R.; Gentilcore, E.; Natale, S.; Cassader, M.; Rizzetto, M.; et al. Plasma adiponectin in nonalcoholic fatty liver is related to hepatic insulin resistance and hepatic fat content, not to liver disease severity. *J. Clin. Endocrinol. Metab.* **2005**, *90*, 3498–3504. [CrossRef]

50. Gastaldelli, A.; Kozakova, M.; Højlund, K.; Flyvbjerg, A.; Favuzzi, A.; Mitrakou, A.; Balkau, B.; RISC Investigators. Fatty liver is associated with insulin resistance, risk of coronary heart disease, and early atherosclerosis in a large european population. *Hepatology* **2009**, *49*, 1537–1544. [CrossRef]

51. Yamauchi, T.; Kamon, J.; Minokoshi, Y.; Ito, Y.; Waki, H.; Uchida, S.; Yamashita, S.; Noda, M.; Kita, S.; Ueki, K.; et al. Adiponectin stimulates glucose utilization and fatty-acid oxidation by activating AMP-activated protein kinase1. *Nat. Med.* **2002**, *8*, 1288–1295. [CrossRef] [PubMed]

52. Zhou, G.; Myers, R.; Li, Y.; Chen, Y.; Shen, X.; Fenyk-Melody, J.; Wu, M.; Ventre, J.; Doebber, T.; Fujii, N.; et al. Role of AMP-activated protein kinase in mechanism of metformin action. *J. Clin. Investig.* **2001**, *108*, 1167–1174. [CrossRef] [PubMed]

53. Jelenik, T.; Rossmeisl, M.; Kuda, O.; Jilkova, Z.M.; Medrikova, D.; Kus, V.; Hensler, M.; Janovska, P.; Miksik, I.; Baranowski, M.; et al. AMP-activated protein kinase {alpha}2 subunit is required for the preservation of hepatic insulin sensitivity by n-3 polyunsaturated fatty acids. *Diabetes* **2010**, *59*, 2737–2746. [CrossRef] [PubMed]

54. Andreelli, F.; Foretz, M.; Knauf, C.; Cani, P.D.; Perrin, C.; Iglesias, M.A.; Pillot, B.; Bado, A.; Tronche, F.; Mithieux, G.; et al. Liver AMPKalpha2 catalytic subunit is a key target for the control of hepatic glucose production by adiponectin and leptin but not by insulin. *Endocrinology* **2006**, *147*, 2432–2441. [CrossRef]

55. Petersen, M.C.; Shulman, G.I. Roles of diacylglycerols and ceramides in hepatic insulin resistance. *Trends Pharmacol. Sci.* **2017**, *38*, 649–665. [CrossRef]

56. Ferramosca, A.; Zara, V. Dietary fat and hepatic lipogenesis: Mitochondrial citrate carrier as a sensor of metabolic changes. *Adv. Nutr.* **2014**, *5*, 217–225. [CrossRef]

57. Sanders, F.W.B.; Acharjee, A.; Walker, C.; Marney, L.; Roberts, L.D.; Imamura, F.; Jenkins, B.; Case, J.; Ray, S.; Virtue, S.; et al. Hepatic steatosis risk is partly driven by increased de novo lipogenesis following carbohydrate consumption. *Genome Biol.* **2018**, *19*, 79. [CrossRef]

58. Berge, R.K.; Madsen, L.; Vaagenes, H.; Tronstad, K.J.; Gottlicher, M.; Rustan, A.C. In contrast with docosahexaenoic acid, eicosapentaenoic acid and hypolipidaemic derivatives decrease hepatic synthesis and secretion of triacylglycerol by d-creased diacylglycerol acyltransferase activity and stimulation of fatty acid oxidation. *Biochem. J.* **1999**, *343 Pt 1*, 191–197. [CrossRef]
59. Kroupova, P.; Van Schothorst, E.; Keijer, J.; Bunschoten, A.; Vodicka, M.; Irodenko, I.; Oseeva, M.; Zacek, P.; Kopecky, J.; Rossmeisl, M.; et al. Omega-3 Phospholipids from krill oil enhance intestinal fatty acid oxidation more effectively than omega-3 triacylglycerols in high-fat diet-fed obese mice. *Nutrients* **2020**, *12*, 2037. [CrossRef]
60. Sharma, L.; Lone, N.A.; Knott, R.M.; Hassan, A.; Abdullah, T. Trigonelline prevents high cholesterol and high fat diet induced hepatic lipid accumulation and lipo-toxicity in C57BL/6J mice, via restoration of hepatic autophagy. *Food Chem. Toxicol.* **2018**, *121*, 283–296. [CrossRef]
61. Zhang, J.; Yang, A.; Wu, Y.; Guan, W.; Xiong, B.; Peng, X.; Wei, X.; Chen, C.; Liu, Z. Stachydrine ameliorates carbon tetrachloride-induced hepatic fibrosis by inhibiting inflammation, oxidative stress and regulating MMPs/TIMPs system in rats. *Biomed. Pharmacother.* **2018**, *97*, 1586–1594. [CrossRef]
62. Tang, W.W.; Wang, Z.; Levison, B.S.; Koeth, R.A.; Britt, E.B.; Fu, X.; Wu, Y.; Hazen, S.L. Intestinal microbial metabolism of phosphatidylcholine and cardiovascular risk. *N. Engl. J. Med.* **2013**, *368*, 1575–1584. [CrossRef]
63. de Souza, C.O.; Teixeira, A.A.S.; Biondo, L.A.; Lima Junior, E.A.; Batatinha, H.A.P.; Rosa Neto, J.C. Palmitoleic acid improves metabolic functions in fatty liver by PPARα-dependent AMPK activation. *J. Cell. Physiol.* **2017**, *232*, 2168–2177. [CrossRef] [PubMed]
64. Liu, G.; Gibson, R.A.; Callahan, D.; Guo, X.-F.; Li, D.; Sinclair, A.J. Pure omega 3 polyunsaturated fatty acids (EPA, DPA or DHA) are associated with increased plasma levels of 3-carboxy-4-methyl-5-propyl-2-furanpropanoic acid (CMPF) in a short-term study in women. *Food Funct.* **2020**, *11*, 2058–2066. [CrossRef] [PubMed]
65. Prentice, K.J.; Wendell, S.G.; Liu, Y.; Eversley, J.A.; Salvatore, S.R.; Mohan, H.; Brandt, S.L.; Adams, A.C.; Serena Wang, X.; Wei, D.; et al. CMPF, a metabolite formed upon prescription omega-3-acid ethyl ester supplementation, prevents and reverses steatosis. *EBioMedicine* **2018**, *27*, 200–213. [CrossRef] [PubMed]
66. Sherriff, J.L.; O'Sullivan, T.A.; Properzi, C.; Oddo, J.-L.; Adams, L.A. Choline. Its potential role in nonalcoholic fatty liver disease, and the case for human and bacterial genes. *Adv. Nutr.* **2016**, *7*, 5–13. [CrossRef] [PubMed]
67. Qian, K.; Zhong, S.; Xie, K.; Yu, D.; Yang, R.; Gong, D.-W. Hepatic ALT isoenzymes are elevated in gluconeogenic conditions including diabetes and suppressed by insulin at the protein level. *Diabetes Metab. Res. Rev.* **2015**, *31*, 562–571. [CrossRef]

Article

A 2 Week Cross-over Intervention with a Low Carbohydrate, High Fat Diet Compared to a High Carbohydrate Diet Attenuates Exercise-Induced Cortisol Response, but Not the Reduction of Exercise Capacity, in Recreational Athletes

Rieneke Terink [1,*], Renger F. Witkamp [1], Maria T. E. Hopman [2], Els Siebelink [1], Huub F. J. Savelkoul [3] and Marco Mensink [1]

1   Division of Human Nutrition and Health, Wageningen University & Research (WUR), 6700 AH Wageningen, The Netherlands; renger.witkamp@wur.nl (R.F.W.); els.siebelink@wur.nl (E.S.); marco.mensink@wur.nl (M.M.)
2   Department of Physiology, Radboud University Nijmegen, 6525 GA Nijmegen, The Netherlands; Maria.Hopman@radboudumc.nl
3   Cell Biology and Immunology Group, Wageningen University & Research (WUR), 6700 AH Wageningen, The Netherlands; huub.savelkoul@wur.nl
*   Correspondence: rieneke.terink@wur.nl or rterink@zgv.nl; Tel.: +31-317-484067

Citation: Terink, R.; Witkamp, R.F.; Hopman, M.T.E.; Siebelink, E.; Savelkoul, H.F.J.; Mensink, M. A 2 Week Cross-over Intervention with a Low Carbohydrate, High Fat Diet Compared to a High Carbohydrate Diet Attenuates Exercise-Induced Cortisol Response, but Not the Reduction of Exercise Capacity, in Recreational Athletes. *Nutrients* **2021**, *13*, 157. https://doi.org/10.3390/nu13010157

Received: 10 December 2020
Accepted: 1 January 2021
Published: 6 January 2021

Publisher's Note: MDPI stays neutral with regard to jurisdictional claims in published maps and institutional affiliations.

Abstract: Low carbohydrate, high fat (LCHF) diets are followed by athletes, but questions remain regarding effects of LCHF on metabolic adaptation, exercise-induced stress, immune function and their time-course. In this cross-over study, 14 recreational male athletes (32.9 ± 8.2 years, VO2max 57.3 ± 5.8 mL/kg/min) followed a two week LCHF diet (<10 En% carbohydrates (CHO), ~75En% Fat) and a two week HC diet (>50 En% CHO), in random order, with a wash-out period of >2 weeks in between. After 2 days and 2 weeks on either diet, participants performed cycle ergometry for 90 min at $60\%W_{max}$. Blood samples for analysis of cortisol, free fatty acids (FFA), glucose and ketones, and saliva samples for immunoglobin A (s-IgA) were collected at different time points before and after exercise. The LCHF diet resulted in higher FFA, higher ketones and lower glucose levels compared to the HC diet ($p < 0.05$). Exercise-induced cortisol response was higher after 2 days on the LCHF diet (822 ± 215 nmol/L) compared to 2 weeks on the LCHF diet (669 ± 243 nmol/L, $p = 0.004$) and compared to both test days following the HC diet (609 ± 208 and 555 ± 173 nmol/L, both $p < 0.001$). Workload was lower, and perceived exertion higher, on the LCHF diet compared to the HC diet on both occasions. A drop in s-IgA following exercise was not seen after 2 days on the LCHF diet, in contrast to the HC diet. In conclusion, the LCHF diet resulted in reduced workload with metabolic effects and a pronounced exercise-induced cortisol response after 2 days. Although indications of adaptation were seen after 2 weeks on the LCHF diet, work output was still lower.

Keywords: cortisol; ketones; s-IgA; exercise; low carbohydrate diet

1. Introduction

Chronic or periodized low carbohydrate, high fat (LCHF) dietary strategies have been applied in sports for several decades. More recently, the interest among some athletes appears to be on the rise again due to the alleged positive effects of 'ketogenic' LCHF (K-LCHF) diets and(or) ketone bodies in general [1]. Typically, K-LCHF diets deliver less than 5% of their energy from carbohydrate (CHO) and more than 75% from fat, corresponding to roughly <50 g/day CHO for most athletes [1]. The more general term (non-ketogenic) LCHF is typically used for diets with <15–20 En% from CHO.

It has been shown that (K-)LCHF diets can increase the transport, uptake and beta-oxidation of fat in muscle. In addition, studies have demonstrated an enhanced activation of some enzymes and mediators involved in adaptation to endurance training during situations of low CHO availability [2–5]. However, studies have also shown that exercise performance, especially at high intensity, is impaired following LCHF diets [6,7]. It is found that LCHF diets can lead to a lower ability to oxidise exogenous CHO during exercise [8], in which a suppression of pyruvate dehydrogenase may play a role [9]. These effects, together with low muscle glycogen stores, result in reduced training intensity when following a LCHF diet, and can attenuate training-induced improvements in exercise performance [10,11].

It has also been suggested that training with low CHO availability may lead to an increased exercise-induced stress response, reflected by higher cortisol levels, and may lead to an attenuated immune response [12]. This might argue against this use of LCHF diets in sports practice, in particular in view of an often already increased risk for upper respiratory tract infections (URTIs) in athletes [13]. The incidence of URTI in athletes is, amongst others, related to low levels of salivary Immunoglobulin A (s-IgA) [13]. This s-IgA is an antibody isotype that is produced locally by B lymphocytes present in mucosal tissues and appears in mucosal secretions such as saliva, thereby protecting against bacteria and viruses entering the body [13]. A shortage of CHO as energy substrate might stimulate cortisol release, inhibiting B-cell immunoglobulin production, resulting in lower s-IgA levels [14]. To our knowledge, only 2 studies investigated the effect of low CHO availability on s-IgA levels [15,16].

When it comes to the time-course of effects of CHO restriction on performance and immune status, several knowledge gaps still exist. This prompted us to carry out the current study in which the short-term stress response following switching to a LCHF diet was compared with the longer-term adaptative response. Therefore, in the present study, we investigated the effect on the exercise-induced cortisol, s-IgA and metabolic responses of acute (2 days) and prolonged (2 weeks) adherence to a LCHF diet, compared to a high CHO (HC) diet. Cortisol, s-IgA levels, upper respiratory tract symptoms (URTS), respiratory exchange ratio (RER), circulating metabolites, work output and perceived exertion during exercise were measured in this randomized cross-over dietary intervention study. We hypothesized that a LCHF diet would result in increased exercise-induced cortisol responses, reduced s-IgA levels, and a reduced work output, whether or not in combination with increased perceived exertion.

2. Materials and Methods

2.1. Participants

A total of fourteen recreational male athletes participated in this study. They were recruited by contacting local cycling and triathlon clubs and via social media. All trained regularly, at least 4 h per week. Additional inclusion criteria were a BMI between 18.5 and 25 kg/m^2 and age between 18 and 45 years. Exclusion criteria were: presence of food allergies, chronic illnesses, use of asthma-, anti-inflammatory- and/or immunosuppressive medication. All participants needed to have a hemoglobin concentration >8.5 mmol/L, and they had not donated blood during six weeks prior to the study.

Study enrolment took place between October 2018 and January 2019. The study was conducted at the Human Nutrition Research Unit, Wageningen University & Research. It was approved by the Medical Ethical Committee of Wageningen University (NL6540408118, ClinicalTrials.gov ID: NCT04019730) and conducted in accordance with the Declaration of Helsinki. All participants gave written informed consent prior to participation.

2.2. Study Design

In a randomized, cross-over design, participants completed two 2-week dietary interventions (Figure 1). General participants characteristics were determined before the start of the first dietary intervention. These included an assessment of maximal aerobic capacity (VO$_{2max}$ test), body composition measurements and questionnaires. Before the intervention, dietary intake was assessed to gain insight in the participants current habitual eating habits and to make an estimation of energy needs. Dietary guidelines were individually explained to participants before the start of each dietary intervention period. Both dietary interventions were followed for 2 weeks. Each dietary intervention period included two exercise test days: one after 2 days on the diet and a second test day after 2 weeks on the diet. The first test day was used to investigate the acute response, and the second test day for the chronic response. Research showed that 5 days adherence to a LCHF diet already resulted in increased fat oxidation [17], therefore we chose to measure after 2 days for a 'stress response' from switching to a LCHF diet. A wash-out period, consisting of their habitual diet, of at least two weeks was applied between both diets. Two weeks seemed long enough, as changing to a LCHF diet already leads to adaptations within 5 days [17], and changing back to a HC diet, leads to 'baseline levels' after again 5 to 6 days [18]. An upper respiratory tract symptoms (URTS) questionnaire was filled out before the intervention and 2 weeks after each diet ended (Figure 1).

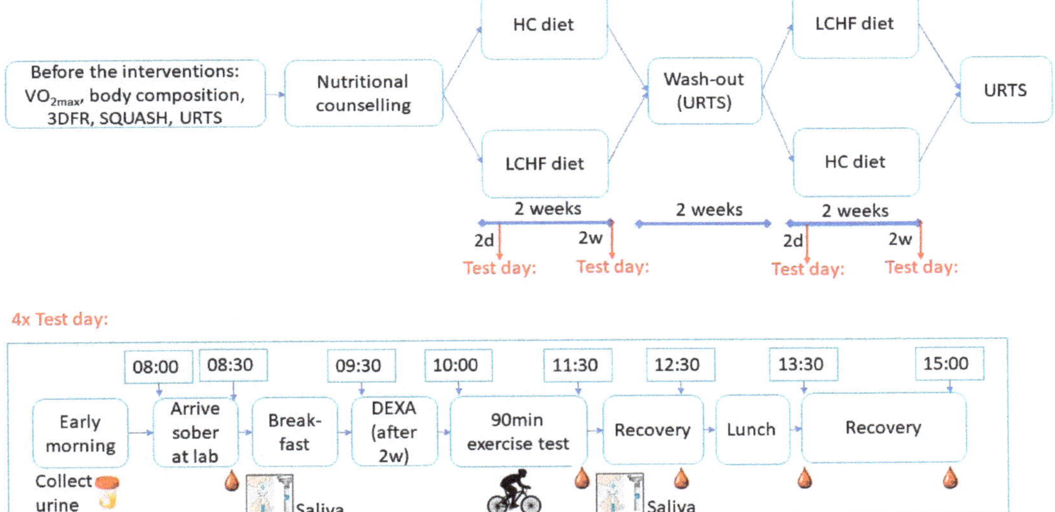

Figure 1. Study design. Schematic study design, 3DRF: 3-day food record; SQUASH: Short Questionnaire to Assess Health enhancing physical activity; URTS: Upper respiratory tract symptoms questionnaire; 2d: after 2 days on the diet; 2w: after 2 weeks on the diet; DEXA: dual energy x-ray absorptiometry.

2.3. Maximal Aerobic Capacity and Body Composition

A maximal exercise test on a bicycle ergometer (Lode Excalibur, Groningen, The Netherlands) was performed to establish maximal aerobic capacity (VO₂max). After an initial workload of 100 Watt for 5 min, workload was subsequently increased by either 25 W/min or 40 W/2 min until the participant could not maintain the required pedaling frequency of at least 60 rpm. Participants were allowed to eat and drink before the test; nothing specific was prescribed. Oxygen consumption was measured with indirect calorimetry (Oxycon Carefusion, Hoechberg, Germany), and VO2 max was recorded [19]. Heart rate was monitored by using a heart rate monitor (Polar T31-coded, Oulu, Finland) and connected exercise tracker (Polar FT1). In addition, body length (Seca 213 portable stadiometer, Hamburg, Germany) and weight (Seca 761 scale) were measured. Thereafter, DEXA measurements were carried out using a Lunar Prodigy Advanced DEXA scanner (GE Health Care, Madison, WI, USA) [20]. A quality assurance test was performed to ensure system suitability and precision of the scanner. Whole body scans were performed according to the manufacturer's protocol and identical scan protocols were used for all subjects.

2.4. Dietary Intervention and Physical Activity

Food diaries were obtained before the intervention using a 3-day food record (3DFR) (2 week days and 1 weekend day, randomly assigned). These were analysed for the total energy intake and macronutrient distribution using Compl-eat software ^{TM} (Department of Human Nutrition and Health, Wageningen University, www.compl-eat.nl) [21]. Personalized diet plans were designed based on the participants estimated total energy needs. In total, 6 energy groups were considered: from 10 to 15 MJ with increments of 1 MJ. Participants were instructed to strictly follow their personalized diets. The diets were either a low carbohydrate, high fat diet aiming for ketogenesis (<10 En% Carbohydrates and ~75 En% Fats) or a high carbohydrate diet (~50 En% Carbohydrates and ~35 En% Fats). Protein intake was supposed to be equal in both diets with 15 En%.

Habitual physical activity was assessed before the start of the intervention, using a questionnaire for physical activity level (Short Questionnaire to Assess Health enhancing physical activity (SQUASH)) [22]. Participants were advised to keep their physical activity level the same during both diets, although this was not tracked with a wearable.

2.5. Nutritional Counselling

Each participant individually received nutritional counselling. A detailed menu for two weeks and some standard products were provided. For the HC diet these were: 30+ cheese (cheese with less fat per 100 gram), sunflower oil, margarine, nuts, muesli bars, fruit juices. For the LCHF diet these comprised: 48+ cheese (cheese with more fat per 100 g), olive oil, margarine, nuts, low-carb bread and beet muffins. The detailed menu consisted of a shopping list, prescribed recipes for breakfast, lunch, dinner and snacks of every day of the week and information about drinks (water, coffee and tea without sugar or milk were allowed) and herbs. Participants received electronically weighing scales (Impuls, Inter-East B.V., Roosendaal, The Netherlands) to precisely measure their dietary intake to ensure that the prescribed menus were followed during the two weeks of intervention. Participants had to weigh all products, except for bread, which was measured in standardized household portion sizes. Deviations from the diet were written down by the participants and leftovers were measured at the end of both intervention periods to assess compliance. Dietary intake was assessed at the end of each diet by calculating the deviations from the diet that were written down by the participants and by subtracting the leftovers from the provided foods which were weighted by distribution and return.

2.6. Exercise Test Days

Test days were performed after 2 days and 2 weeks on each of the diets. See Figure 1 for an overview of the test day. Participants arrived after an overnight fast. At home, they already collected their morning urine to assess ketosis (ketostick, strips 50 A2880 B51, Bayer, Leverkusen, Germany). At 08:00 AM an intravenous cannula was inserted in an antecubital vein and a first blood sample was taken at 08:30 AM. Simultaneously, participants donated saliva via unstimulated, passive drool [23]. A standardized breakfast customized to their energy needs and current diet was provided after the first blood drawing (LCHF breakfast: 588 kCal (average) ~74 En% fat, 17 En% protein, 6 En% carb; HC breakfast: 505 kCal (average) ~34 En% fat, 15 En% protein, 48 En% carb). Thereafter, only after 2 weeks on both diets, body composition was assessed using a DEXA scan (GE Health Care, Madison, WI, USA). Scans were performed on the same time of the day during all sessions to minimize measurement errors.

Next, a 90 min bicycle ergometer test (Lode Excalibur, Groningen, The Netherlands) at 60% of the athletes' individual W_{max} (~70% VO_{2max}) was performed from 10:00 a.m. to 11:30 a.m. If an athlete failed to maintain the prescribed workload, the workload was decreased to a level at which the athlete could keep on cycling until the end of the test.

Adjustments were written down and the power multiplied by time in seconds was used to calculate total workload. Workload during the 90 min exercise tests was assessed as area under the curve in kilo joule (kJ).

Heart rate was measured with a heart rate belt (Polar T31-coded, Oulu, Finland), placed around the chest. Gaseous exchange was measured (Oxycon Carefusion, Hoechberg, Germany) before the start of the exercise test (while sitting still on the bike for 5 min) and at 60 min during the exercise test during a 5 min period, to assess respiratory exchange rate (RER: ratio VCO_2/VO_2).

Participants were allowed to drink plain water during the test, but were not allowed to eat. Drinking was not allowed in the last 10 min of the exercise test. Directly after the exercise test a Borg scale was shown to ask for the rate of perceived exertion (RPE) and another blood sample and saliva sample were taken. Thereafter, participants could take a shower and relax. Another blood sample was taken 1 h after the end of the exercise. Subsequently participants received a standardized lunch customized to their energy needs and current diet (LCHF lunch: 1027 kCal (average) ~78 En% fat, 16 En% protein, 4 En% carb; HC lunch: 781 kCal (average) ~31 En% fat, 14 En% protein, 52 En% carb). Two more blood samples were taken at 2 and 3.5 h after exercise, respectively. These time points were chosen because we aimed to analyse the immunological response more in depth at a later stage. This would include, for example, cytokine responses, and based on their different reaction times, we collected samples at these time points (Figure 1).

2.7. Blood Sampling and Analysis

Blood samples were collected in lithium-heparin, EDTA and serum tubes. Lithium-heparin tubes (4.5 mL LH PST™ II, Becton-Dickinson, New Jersey, America) were centrifuged at 1300 CRF for 10 min at room temperature (RT), plasma was frozen at $-80\,^{\circ}C$ until it was analyzed for glucose concentrations. Glucose was measured by means of an end-point technique (Siemens, The Netherlands). EDTA tubes (8 mL, Becton-Dickinson, New Jersey, America) were centrifuged at 1200 G for 15 min at 4 $^{\circ}C$, and plasma was frozen at -80 degrees until it was analysed for free fatty acids concentrations. Free fatty acids were assessed using an enzymatic test kit according to the manufacturer's protocol (InstruChemie, Delfzijl, The Netherlands). Serum tubes (5 mL, Becton-Dickinson, NJ, USA) were set aside for at least 30 min, where after they were centrifuged at 1300 G for 10 min at RT, serum was frozen at -80 degrees until it was analysed for ketone content and cortisol concentration. Beta-hydroxybutyrate (βHB) was determined via a colorimetric enzymatic assay (Sigma-Aldrich; St. Louis, MO, USA). Analysis was performed according to manufacturer's protocol. Cortisol was measured with immunometric chemiluminescence (sandwich) assay with Immulite XPi (Siemens, Den Haag, The Netherlands).

2.8. Saliva Sampling and Analysis

Saliva was collected at two time points at every test day: one before breakfast and one directly after exercise. In order to collect whole saliva from the mouth, unstimulated, passive drool was performed [23]. Participants were asked to bend their head slightly downwards and first collect some saliva in their mouth before drooling into the saliva collection aid (Salimetrics, LLC, State College, PA, USA). At least 0.5 mL of saliva was collected in 2-mL collection tubes (Wheaton, Millville, NJ, USA) per time point. Samples were temporarily stored on dry ice and transferred to a refrigerator at $-80\,^{\circ}C$ within seven hours until analysis. IgA antibodies in saliva were determined by enzyme-linked immunosorbent assay (ELISA) as described before [24]. The samples for each individual participant were run on the same assay to eliminate inter-assay variance.

2.9. URTS Questionnaires

Before the intervention, and two weeks after the final day of each dietary intervention, participants received a questionnaire about symptoms related to upper respiratory tract infections (URTI). This questionnaire was a Dutch translation of the validated WURSS-21 questionnaire [25].

2.10. Statistical Analysis

Data was analysed using IBM SPSS version 27 Statistical Package for Social Sciences (IBM SPSS version 27.0, Armonk, New York, NY, USA). Except for URTS, all data were normally distributed. A paired t-test was performed to assess differences between the LCHF and HC diets. A two-way repeated measures ANOVA (two factor, time x diet) was performed to analyse work, RER, HR and Rate of Perceived Exertion, s-IgA and the cortisol and metabolic response to both diets. When an effect of condition or time or interaction was identified, a pairwise multiple comparison with Bonferroni correction was done to identify the differences. URTS data was analysed using a sign test and the correlation between URTS and s-IgA data was performed using a Spearman correlation test, as data were not normally distributed. The level of significance was set at $p < 0.05$. Data are presented as mean $\pm$ SD unless indicated otherwise.

3. Results

3.1. Participant Characteristics

Baseline characteristics of the fourteen participants are depicted in Table 1. They were active in a variety of sports (cyclist ($n = 5$), triathlete ($n = 1$), climber ($n = 2$), strength trainer ($n = 2$), swimmer ($n = 1$), volleyball player ($n = 1$), football player ($n = 1$), runner ($n = 1$)). They were 32.9 $\pm$ 8.2 years old and had a VO2max of 57.3 $\pm$ 5.8 mL/kg/min. Their habitual diet contained 2961 $\pm$ 528 kCal, 36 $\pm$ 6 En% fat, 16 $\pm$ 3 En% protein, 43 $\pm$ 5 En% carbs. Their habitual training consisted of 5.6 $\pm$ 1.1 training hours per week.

Table 1. Participant characteristics.

	Participants ($n = 14$)
Age (years)	32.9 $\pm$ 8.2
Body composition	
Height (cm)	181.7 $\pm$ 4.7
Weight (kg)	76.4 $\pm$ 5.4
BMI (kg/m^2)	23.1 $\pm$ 1.4
Lean mass (kg)	61.9 $\pm$ 3.4
Lean mass (%)	81.3 $\pm$ 4.4
BMC (kg)	3.2 $\pm$ 0.25
BMC (%)	4.2 $\pm$ 0.32
Body fat (kg)	11.2 $\pm$ 4.0
Body fat (%)	14.5 $\pm$ 4.6
Total training (hours/week)	5.6 $\pm$ 1.1
Maximal exercise performance	
VO2max (ml/kg/min)	57.3 $\pm$ 5.8
Max heart rate (bpm)	187 $\pm$ 9
Max Power (Watt)	346 $\pm$ 46
Max Power/kg body weight	4.5 $\pm$ 0.5

Means $\pm$ SD are shown. BMI: Body mass index; BMC: Bone mineral content. Physical characteristics are determined during a VO2max test.

3.2. Dietary Intake and Blood and Urine Ketone Levels

Energy intake between the LCHF (3104 $\pm$ 297 kCal) and the HC diet (3075 $\pm$ 298 kCal) was not different ($p = 0.221$). As intended, macronutrient intake was significantly different between both diets, with significantly higher fat intake in the LCHF diet compared to the HC diet (73 $\pm$ 1 vs. 33 $\pm$ 0 En%, for LCHF and HC, respectively; $p < 0.001$) and, in line with the experimental design, a lower carbohydrate intake in the LCHF diet compared to the HC diet (8 $\pm$ 0 vs. 49 $\pm$ 0 En%, for LCHF and HC, respectively; $p < 0.001$). Protein intake was higher in the LCHF diet compared to the HC diet (16 $\pm$ 1 vs. 15 $\pm$ 0 En%, for LCHF and HC, respectively; $p < 0.001$), although this was not intended. An overview of the total daily energy intake and macronutrient distribution at baseline and during both dietary intervention periods, can be seen in Table 2.

Table 2. Dietary intake and fasting serum and urine ketone levels.

	Habitual	LCHF Diet	HC Diet	*p* Value
Energy (kCal)	2961 ± 528	3104 ± 297	3075 ± 298	0.221
Protein (g/day)	116 ± 22	124 ± 12	112 ± 11	<0.001
Protein (En%)	16 ± 3	16 ± 1	15 ± 0	<0.001
Carbohydrate (g/day)	318 ± 72	64 ± 6	373 ± 38	<0.001
Carbohydrate (En%)	43.4 ± 5.3	8 ± 0	49 ± 0	<0.001
Total Fat (g/day)	122 ± 29	254 ± 25	116 ± 11	<0.001
Total Fat (En%)	36 ± 6	73 ± 1	33 ± 0	<0.001
Saturated Fat (g/day)	43 ± 13	68 ± 6	32 ± 3	<0.001
Saturated Fat (En%)	13.1 ± 3.2	19.7 ± 0.3	9.3 ± 0.3	<0.001
Monounsaturated Fat (g/day)	46 ± 13	127 ± 13	35 ± 3	<0.001
Monounsaturated Fat (En%)	13.9 ± 3.3	36.8 ± 1.1	10.3 ± 0.2	<0.001
Polyunsaturated Fat (g/day)	22 ± 7	39 ± 5	41 ± 5	0.002
Polyunsaturated Fat (En%)	6.6 ± 1.8	11.4 ± 0.4	12.1 ± 0.2	<0.001
Cholesterol (mg/day)	354 ± 242	699 ± 57	165 ± 18	<0.001
Dietary Fiber (g/day)	31 ± 6	28 ± 3	41 ± 4	<0.001
Dietary Fiber (En%)	2 ± 0	5 ± 1	9 ± 2	<0.001
Fasting serum βHB (mmol/L)		0.27 ± 0.13	0.07 ± 0.04	<0.001
Urine ketone levels (g/L)		0.26 ± 0.25	0.00 ± 0.00	<0.001

Means ± SD are shown. *p*-values represent a dependent t-test between both intervention diets (LCHF vs. HC); serum βHB and urine ketone bodies represent data after following the diets for 2 weeks.

The LCHF diet was geared to induce nutritional ketosis. Deviations from the prescribed diets were negligible. Urine ketone levels ranged from 0–1.6 g/L (average: 0.16 ± 0.42 g/L) after 2 days on the LCHF diet and ranged from 0–0.8 g/L (0.26 ± 0.25 g/L) after 2 weeks on the LCHF diet. One out of the 14 participant had no detectable ketones in his urine after 2 weeks on the LCHF diet. There were no ketones present in urine samples during the HC diet. Baseline blood ketone (β-hydroxy-butyrate) levels ranged from 0.06–0.68 mmol/L (average: 0.31 ± 0.18 mmol/L) after 2 days and from 0.21–0.97 mmol/L (0.54 ± 0.26 mmol/L) after 2 weeks on the LCHF diet. On the HC diet, baseline ketone levels were significantly lower: after 2 days ranging from 0.06–0.45 mmol/L (0.14 ± 0.10 mmol/L) and after 2 weeks ranging from 0.05–0.32 mmol/L (0.13 ± 0.08 mmol/L) ($p < 0.001$ compared to the LCHF diet for both test days).

3.3. Body Composition

Compared to baseline (76.4 ± 5.4 kg), body mass was significantly lower after 2 weeks on the LCHF diet (74.0 ± 4.5 kg, $p < 0.001$) and after 2 weeks on the HC diet (75.1 ± 4.7 kg, $p = 0.003$). Body mass was also significantly lower after 2 weeks on the LCHF diet compared to body mass after 2 weeks on the HC ($p = 0.005$). Body fat percentage was lower after each of the diets (LCHF: 12.9 ± 4.3% and HC: 13.5 ± 4.6%) compared to baseline (14.5 ± 4.6%, both $p < 0.001$). Body fat percentage was not different between diets ($p = 0.101$). Lean body mass percentage was higher after both diets (LCHF: 82.8 ± 4.2 and HC: 82.2 ± 4.5%) compared to baseline (81.3 ± 4.4%, $p = 0.017$ and $p = 0.011$, respectively). Bone mineral content (BMC) was 4.3 ± 0.3% (3.2 ± 0.2 kg) and comparable between diets ($p = 0.271$). The difference in lean mass percentage between diets was also not significant ($p = 0.110$).

3.4. Work output, Respiratory Exchange Ratio and Perceived Exertion

Exercise data can be found in Table 3. The total work in kJ that had to be performed during the 90 min exercise was 1120 ± 148 kJ. However, exercise intensity had to be reduced on multiple occasions. The total work output was significantly lower during the LCHF diet compared to the HC diet, both after 2 days as well as after 2 weeks (939 ± 163 vs. 1042 ± 151 kJ after 2 days and 1003 ± 129 kJ vs. 1043 ± 141 kJ after 2 weeks, for LCHF and

HC diet, respectively, $p < 0.02$ between diets). Total workload significantly increased on the LCHF diet after 2 weeks compared to 2 days ($p = 0.03$), while no time-effect was seen for the HC diet. Substrate oxidation patterns at rest and during exercise were significantly different between diets. At rest, RER was significantly lower after 2 days and after 2 weeks on the LCHF diet (0.76 ± 0.03 and 0.77 ± 0.06) compared to the HC diet (0.86 ± 0.05 and 0.87 ± 0.05) (both $p < 0.001$). Additionally, during exercise, RER was significantly lower after 2 days and 2 weeks on the LCHF diet (0.82 ± 0.03 and 0.82 ± 0.03) compared to the HC diet (0.90 ± 0.04 and 0.91 ± 0.02) (both $p < 0.001$). Within each diet group, RER at rest and during exercise did not differ between 2 days and 2 weeks ($p > 0.05$). Heart rate during exercise was significantly higher after 2 weeks on the LCHF diet compared to the HC diet (170 ± 11 bpm vs. 165 ± 13 bpm, $p = 0.001$). There was no significant difference in heart rate between the diets after 2 days (165 ± 13 for LCHF vs. 164 ± 18 for HC, $p = 0.652$). Participants rated their perceived exertion higher after 2 days on the LCHF diet compared to 2 days on the HC diet (18.0 ± 1.4 vs. 15.5 ± 2.7, for LCHF vs. HC; $p = 0.001$). This difference in perceived exertion diminished after 2 weeks, but still tended to be higher on the LCHF diet (17.3 ± 1.7 vs. 16.1 ± 2.0, for LCHF and HC; $p = 0.053$).

Table 3. Work, RER, HR and RPE.

	LCHF			HC			Intervention	
			Time Effect			Time Effect	after 2d	after 2w
	after 2 Days	after 2 Weeks	p-Value	after 2 Days	after 2 Weeks	p-Value	p-Value	p-Value
Work (AUC in kJ)	939 ± 163	1003 ± 129	0.030	1042 ± 151	1043 ± 141	0.974	0.004	0.016
RER (rest)	0.76 ± 0.03	0.77 ± 0.06	0.282	0.86 ± 0.05	0.87 ± 0.05	0.564	<0.001	<0.001
RER (at t60)	0.82 ± 0.03	0.82 ± 0.03	0.681	0.90 ± 0.04	0.91 ± 0.04	0.612	<0.001	<0.001
HR (in bpm; at t60)	165 ± 13	170 ± 11	0.014	164 ± 18	165 ± 13	0.633	0.652	0.001
RPE score	18.0 ± 1.4	17.3 ± 1.7	0.151	15.5 ± 2.7	16.1 ± 2.0	0.300	0.001	0.053

Values are mean $\pm$ SD, calculated after 2 days and 2 weeks on both diets. LCHF: low carbohydrate high fat diet; HC: high carbohydrate diet; AUC: area under the curve; kJ: kilo Joule; RER: respiratory exchange ratio; t60: after 60 min exercise; HR: heart rate; RPE: rate of perceived exertion. p-values represent repeated measures ANOVA.

3.5. Blood Metabolites (Free Fatty Acids, Glucose, Cortisol and Ketone Bodies)

Blood metabolite levels over time are depicted in Figure 2. Circulating markers of lipid metabolism indicated a significantly difference between the HC and LCHF diet. Serum free fatty acids (FFAs) at baseline were comparable between diets and test days ($p > 0.05$). However, peak FFAs levels at the end of the exercise were significantly higher with the LCHF diet (3.4 ± 0.9 and 3.7 ± 0.8 mmol/L after 2 days and 2 weeks, respectively) compared to the HC diet group (2.3 ± 0.6 and 2.2 ± 0.5 mmol/L after 2 days and 2 weeks, respectively, $p < 0.001$ vs. LCHF). Serum beta-Hydroxy-Butyrate (β-HB) levels were significantly higher with the LCHF diet compared to those with the HC diet at all time points, and at both test days ($p < 0.001$).

Glucose levels were in general lower on the LCHF diet compared to those on the HC diet. Baseline glucose levels were not different between diets after 2 days on each of the diets (4.7 ± 0.6 vs. 4.9 ± 0.4 mmol/L for LCHF vs. HC, $p = 0.153$), but were significantly lower after 2 weeks on the LCHF diet (4.7 ± 0.4 vs. 5.0 ± 0.4 mmol/L, for LCHF vs. HC; $p = 0.035$). The exercise-induced decrease in glucose was large on the LCHF diet (-1.00 ± 0.76 mmol after 2 days and -0.76 ± 0.27 mmol/L after 2 weeks, both $p < 0.001$ compared to baseline glucose levels), and much smaller after 2 days on the HC diet (-0.26 ± 0.37 mmol/L, $p = 0.018$ compared to baseline glucose levels) or even absent after 2 weeks on the HC diet (-0.018 ± 0.49 mmol/L, $p = 0.192$). After lunch (2 h after exercise), glucose levels increased with both diets, but to a greater extent on the HC diet (Figure 2C).

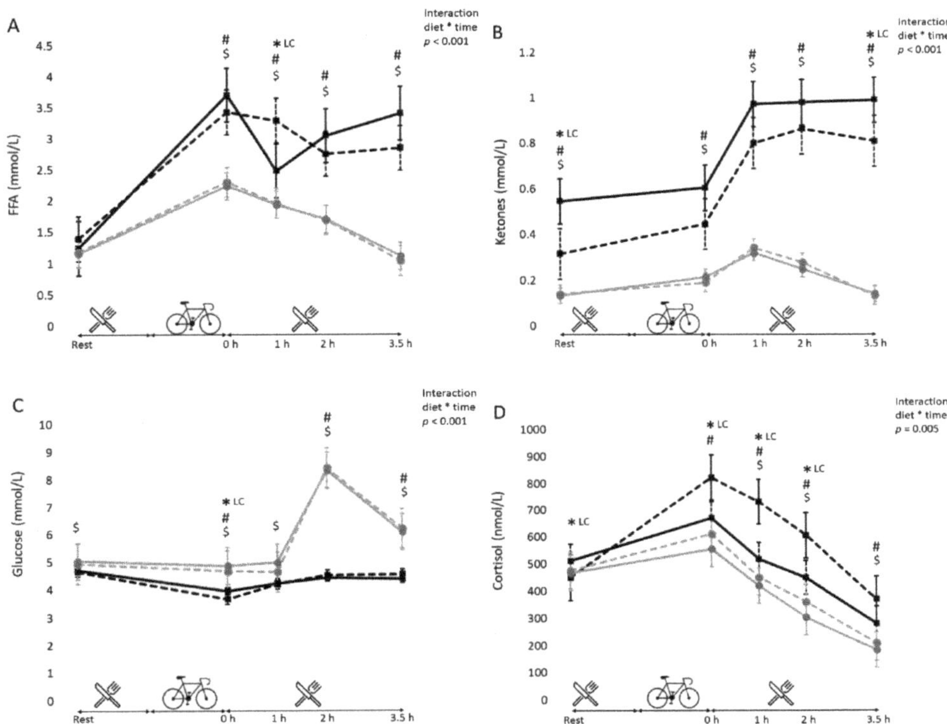

Figure 2. Metabolites. Circulating concentrations of Free fatty acids (**A**), Ketones (**B**), Glucose (**C**) and Cortisol (**D**) measured after 2 days on the LCHF diet (black dotted lines) and HC diet (grey dotted lines) and after 2 weeks on the LCHF diet (black continues line) and HC diet (grey continues line). Means ± SE are shown. All variables showed significant interactions (diet x time) effects. * LC indicates that this difference was between 2 days and 2 weeks on the LCHF diet. Within the HC diet there were no differences between concentrations after 2 days and 2 weeks on that diet. # indicates significant differences between the LCHF and HC diet after 2 days. $ indicates significant differences between the LCHF and HC diet after 2 weeks.

The exercise induced cortisol response was highest after 2 days on the LCHF diet compared to 2 weeks on the LCHF diet (822 ± 215 nmol/L vs. 669 ± 243 nmol/L, for 2 days vs. 2 weeks; $p = 0.004$) and compared to the HC diet (609 ± 208 and 555 ± 173 nmol/L, for 2 days and 2 weeks on the HC diet, both $p < 0.001$ vs. LCHF diet). After 2 days on the LCHF diet, cortisol levels increased by 83% post-exercise, compared to only a 31% increase after 2 weeks. On the HC diet, this increase was 28 and 19% after 2 days and 2 weeks of diet intervention, respectively. Resting plasma cortisol concentration was not affected by diet, as there were no differences between baseline cortisol levels between the diets after 2 days and between the diets after 2 weeks. See Figure 2D.

3.6. Salivary IgA

No clear exercise effect was seen for salivary IgA1 and IgA2, neither during the HC diet, nor the LCHF diet. See Figure 3. There was a significant interaction between diet and time point (before vs. after exercise) for s-IgA2 after 2 weeks on both diets ($p = 0.049$). Post-exercise s-IgA1 and s-IgA2 levels were lower on the HC diet compared to the LCHF diet after two days adaptation (s-IgA1: 326 ± 143 vs. 502 ± 247 µg/mL for HC and LCHF; $p = 0.004$; s-IgA2: 102 ± 96 vs. 149 ± 162 µg/mL for HC and LCHF; $p = 0.04$).

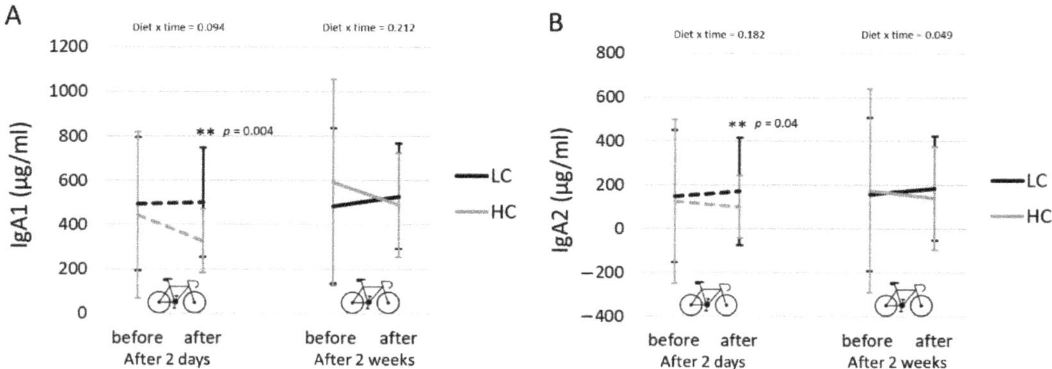

Figure 3. Salivary IgA levels. Salivary IgA1 (**A**) and IgA2 (**B**) levels (mean ± SD) before and after exercise, measured after 2 days on the LCHF diet (black dotted lines) and HC diet (grey dotted lines) and after 2 weeks on the LCHF diet (black continues line) and HC diet (grey continues line). Significant *p* values represent a paired samples t-test for the HC vs. LCHF diet.

3.7. URTS

The URTS scores for the LCHF and HC diet were 1.8 ± 2.3 and 3.6 ± 5.5, respectively. A sign test did not show any statistically significant difference between the two median URTS scores ($p = 0.187$). After both diets, all participants rated the question "how ill do you feel today?" with a 0 "not ill" or a 1 "very mildly" on a scale of 0 to 7 "very ill". For the LCHF diet, only one participant rated this question with a 1, all other participants rated this question with a 0. For the HC diet, 3 participants rated this question with a 1, all others with a 0. In addition, there were no significant correlations between URTS and s-IgA levels, $p > 0.05$.

4. Discussion

We aimed to investigate short-term (2 days) and prolonged (2 weeks) effects of adherence to a LCHF diet with regard to its effects on exercise-induced cortisol, s-IgA and metabolic responses, and compared this to a HC diet. This is because, to our knowledge, studies of the time course of effects after a low carbohydrate diet are relatively scarce, but relevant. We showed that the LCHF diet resulted in a reduced work output and a higher perceived exertion both after 2 days and 2 weeks, in addition to marked metabolic differences and a pronounced exercise-induced cortisol response after 2 days.

Metabolic effects, work output and perceived exertion: Two days on the LCHF or HC diet resulted in different metabolic effects during the exercise trial. After cycling, participants following the LCHF diet showed higher circulating FFA and ketone levels, whereas their plasma glucose levels and RERs were lower at that time point, indicating more reliance on fat oxidation in comparison to the HC diet. Except for a further increase in ketone levels with the LCHF diet, these differences were similar after 2 weeks on the different diets. Plasma free fatty acids at baseline were not different between diets and between days, which can be explained by a lower release of FFA from the liver and adipose tissue during the LCHF diet. Free fatty acids peaked after exercise in the LCHF diet. Plasma FFAs also increased after exercise in the HC diet, which is in agreement with other studies [6,26]. During exercise, the rate of lipolysis increases and, as a result, the concentration of free fatty acids in plasma increases [27]. The higher plasma ketone levels at baseline in the LCHF diet confirm limited CHO availability after 2 days. Apparently, this was not visible yet from blood glucose levels, which were only lower after 2 weeks on the LCHF diet. A decrease of blood glucose levels after exercise following a LCHF diet has also been observed in other studies [6]. The marked glucose peak observed in the HC diet group after consuming the standardized meal was expected as this meal contained ~200 g of carbohydrates.

The reduced work output after following the LCHF diet for 2 days is in line with previous studies and can be explained by decreased CHO oxidation, even though fat oxidation rates may already be increased after short term adaptation [18]. This was emphasized by the higher rates of perceived exertion. Although after 2 weeks on the LCHF diet the work output was higher than after 2 days, suggesting some adaptation towards improved fat oxidation, it was still lower compared to that following the HC diet. At the same time, heart rate was higher and perceived exertion equal between both test days on the LCHF diet. The lower RER after 2 days on the LCHF diet observed in our study is in agreement with other findings suggesting that increased fat oxidation can occur within days during low carbohydrate availability [18,28]. Studies suggest that prolonged adherence to a LCHF diet enhances the breakdown, transport, and oxidation of fat in skeletal muscle [29]. However, this was not reflected by a further decrease of the RER in our study. It should be noted that the interpretation of RER, VO2 and VCO2 values for fat and glucose oxidation requires some caution, as the oxidation of ketone bodies confounds the results [30]. Given the higher ketone levels after 2 weeks on the LCHF diet and the slightly improved work output and lower RPE this might play a role in our study as well. It remains speculative whether longer adherence to the LCHF diet would have resulted in smaller differences in work output with those after the HC diet. It has been reported that adaptation to a non-ketogenic low carbohydrate diet would be around 5 days, without further enhancement thereafter [17]. Others have suggested that consumption of a LCHF diet results in adaptations of the homeostatic regulation of muscle glycogen and even further improved fat oxidation during exercise on a longer term [26].

Cortisol levels: Baseline cortisol levels were comparable between both diets, which is in agreement with a study showing no association between resting cortisol levels and any dietary parameter [31]. The marked effect on the exercise-induced cortisol response after 2 days of following a LCHF diet is likely related to the low CHO availability, which is only partly compensated by increased use of fat as a substrate. In that situation, exercise will rapidly lead to glycogen depletion [32] resulting among others in increased cortisol release [33,34]. The exercise intensity in our study was fairly high, reflected by heart rates above 86% HRmax. Several studies have shown that when individuals perform exercise after several days on very low carbohydrate diets, cortisol levels are markedly higher than with a normal or high carbohydrate diet [35,36]. Interestingly, we found that this difference in cortisol response, compared to after a HC diet, was diminished after two weeks. This suggests further adaptation to the LCHF diet.

s-IgA levels: Salivary IgA1 and IgA2 were both lower post-exercise after 2 days on the HC diet compared to levels after 2 days on the LCHF diet, which might suggest that a short-term LCHF diet attenuates exercise-induced decreases in s-IgA. The reduced s-IgA (both in IgA1 and IgA2) is often linked to an increased risk for URTI [37], despite being also debated [38]. There were no differences in post-exercise s-IgA levels after two weeks with both diets, which is in agreement with a 3-week trial, showing that post-exercise changes in s-IgA were comparable between a HC and ketogenic diet [15]. On the other hand, 70% higher s-IgA secretion rates were reported after a 31-day ketogenic diet compared to secretion rates before this ketogenic diet [16]. On beforehand, we expected lower s-IgA levels with the LCHF diet, because higher cortisol levels can result in lower immunoglobulin production by B-cells, thereby attenuating the s-IgA levels [14]. Some researchers suggest that other factors besides cortisol affect s-IgA levels after exercise, for example increased sympathetic nervous system innervation of the salivary glands or total energy availability [16]. In addition, variation in s-IgA levels was very large between our participants, which can be explained by variation in the health status of the oral cavity [38], as well as by variation in sleep practices, psychological stress and flow rate. Unfortunately, we did not assess salivary flow rate, but only s-IgA concentrations, so direct comparisons with these findings are not possible.

In the majority of studies, no separate detection of s-IgA1 and s-IgA2 levels was performed and given the different susceptibility of these isotypes for proteolysis, this might affect the association with exercise-induced changes in mucosal immunity. Both our findings on s-IgA1 and s-IgA2 connected to this apparent discrepancy with cortisol levels merit further investigation, preferably in a long-term study.

Our comparable s-IgA levels after 2 weeks of LCHF and HC diet were also reflected by finding no differences in URTS in our athletes 2 weeks after the end of each of the diets whereby none of the participants indicated to feel ill. Although it should be noted that these URTS questionnaires are filled in by the participants and not established by an additional throat swab. Unfortunately, the data in this study does not suggest that one of the diets could protect against URTS, although the inhibition of the s-IgA decrease after exercise on the LCHF diet after 2 days seems promising. Several articles already stated that there was no evidence of a beneficial effect of carbohydrates on URTS [39,40]. However, whether a LCHF ketogenic diet would have beneficial effects should be studied in future.

Body mass: Mean body mass was lower after following both diets compared to before the intervention. However, these decreases in body mass were expected after the LC diet [41,42], as a LC diet decreases glycogen concentrations, which is associated with a loss of body water and thus body weight [43]. Decreases on the HC diet were not directly foreseen, but may have been caused by underreported energy intakes in the 3-day food records at intake, leading to a dietary advice with a lower energy intake then needed for the participant. This would account for both the HC diet and the LC diet, as athletes were subscribed the same energy group with both diets. This underreporting is common in the athletic population [44]. We don't think that these changes have affected our results significantly.

Limitations and strengths: To our knowledge, this is the first study applying a cross-over design in which effects of implementing a LCHF diet were measured after 2 days and 2 weeks adherence to the diet, which not only enabled us to compare effects of a LCHF diet with those of a HC diet, but also gain insight in the time-course of these effects.

A limitation is that we did not take salivary flow rate and salivary volume into account, but only salivary IgA concentrations. Although salivary IgA concentrations alone also provide relevant insight, this information would have added to the study since when flow rate and/or volume is low, this might impair mucosal immunity as well.

For some participants, we had to reduce the workload during the exercise sessions, especially during the LCHF test days. This led to a lower work output when participants were on the LCHF diet, which in turn makes it harder to directly compare exercise-induced cortisol responses, metabolites and s-IgA between the diets. On the other hand, perceived exertion and heart rate were higher during the exercise tests during LCHF intervention, which made us conclude that the effort was higher when on the LCHF diet.

In addition, it would have been useful to have additional baseline data for body composition before the start of the second dietary intervention. However, in order to limit the burden of the participants, we had to make decisions on which measurements to include and exclude.

The wash-out period of 2 weeks between the interventions was based on previous findings regarding adaptation time to a LCHF diet [17] and turning back to baseline after CHO loading [18]. Further studies are warranted to explore the time-course of this reversal of effects. In addition, diets were not controlled during the wash-out period, which could have affected our results on the first test day in the second diet. The same holds true for the dietary habits of the participants prior to our study. It has been found that regular consumption of specific food products is associated with differences in exercise-induced muscle damage and cardiac stress [45], which might have influenced cortisol response and perceived exhaustion. Next to this, it is recommendable to repeat the study in female athletes, as they are underreported in research.

In conclusion, the results of the present study showed that 2 days adherence to a LCHF diet already leads to metabolic changes, as reflected by lower RER, lower glucose, higher FFA and higher ketone levels. These metabolic changes were comparable between 2 days and 2 weeks adherence to the LCHF diet, except for ketone levels which were further increased after 2 weeks. On the other hand, the exercise-induced cortisol response was higher after 2 days and attenuated after 2 weeks. The exercise capacity after adherence to the LCHF diet was low, with lower workload, higher or comparable HR and higher RPE compared to the HC diet. A drop in s-IgA following exercise was not seen after 2 days on the LCHF diet, in contrast to the HC diet, which might suggest some form of protective effect, although we could not relate this to URTS. Our results underline that adaptation to a LCHF diet in terms of the metabolic and exercise-cortisol response have different time spans.

Author Contributions: The study was designed by R.T., R.F.W., H.F.J.S. and M.M.; dietary guidelines were designed by E.S.; data were collected and analysed by R.T.; data interpretation and manuscript preparation were undertaken by R.T., R.F.W., H.F.J.S., M.T.E.H., E.S. and M.M. All authors approved the final version of the paper. All authors agree to be accountable for all aspects of the work in ensuring that questions related to the accuracy or integrity of any part of the work are appropriately investigated and resolved. All persons designated as authors qualify for authorship, and all those who qualify for authorship are listed. All authors have read and agreed to the published version of the manuscript.

Funding: This study was part of the EAT2MOVE project and supported by a grant from the Province of Gelderland, proposal PS2014-49.

Declaration: The results of the present study do not constitute endorsement by ACSM. We declare that the results of this study are presented clearly, honestly, and without fabrication, falsification, or inappropriate data manipulation.

Institutional Review Board Statement: The study was conducted according to the guidelines of the Declaration of Helsinki, and approved by the Medical Ethical Committee of Wageningen University (protocol code: NL6540408118, date of approval: 13-07-2018, ClinicalTrials.gov ID: NCT04019730).

Informed Consent Statement: Informed consent was obtained from all subjects involved in the study.

Data Availability Statement: The data presented in this study are available on request from the corresponding author. The data are not publicly available due to ethical reasons.

Conflicts of Interest: There are no conflicts of interest to acknowledge.

References

1. Burke, L.M. Ketogenic low-CHO, high-fat diet: The future of elite endurance sport? *J. Physiol.* **2020**. [CrossRef]
2. Bartlett, J.D.; Hwa Joo, C.; Jeong, T.S.; Louhelainen, J.; Cochran, A.J.; Gibala, M.J.; Morton, J.P. Matched work high-intensity interval and continuous running induce similar increases in PGC-1alpha mRNA, AMPK, p38, and p53 phosphorylation in human skeletal muscle. *J. Appl. Physiol.* **2012**, *112*, 1135–1143. [CrossRef]
3. Cochran, A.J.; Little, J.P.; Tarnopolsky, M.A.; Gibala, M.J. Carbohydrate feeding during recovery alters the skeletal muscle metabolic response to repeated sessions of high-intensity interval exercise in humans. *J. Appl. Physiol.* **2010**, *108*, 628–636. [CrossRef] [PubMed]
4. Sanders, M.J.; Grondin, P.O.; Hegarty, B.D.; Snowden, M.A.; Carling, D. Investigating the mechanism for AMP activation of the AMP-activated protein kinase cascade. *Biochem. J.* **2007**, *403*, 139–148. [CrossRef] [PubMed]
5. Wojtaszewski, J.F.; MacDonald, C.; Nielsen, J.N.; Hellsten, Y.; Hardie, D.G.; Kemp, B.E.; Richter, E.A. Regulation of 5′AMP-activated protein kinase activity and substrate utilization in exercising human skeletal muscle. *Am. J. Physiol. Endocrinol. Metab.* **2003**, *284*, E813–E822. [CrossRef] [PubMed]
6. Burke, L.M.; Angus, D.J.; Cox, G.R.; Cummings, N.K.; Febbraio, M.A.; Gawthorn, K.; Hargreaves, M. Effect of fat adaptation and carbohydrate restoration on metabolism and performance during prolonged cycling. *J. Appl. Physiol.* **2000**, *89*, 2413–2421. [CrossRef] [PubMed]
7. Burke, L.M.; Sharma, A.P.; Heikura, I.A.; Forbes, S.F.; Holloway, M.; McKay, A.K.; Bone, J.L.; Leckey, J.J.; Welvaert, M.; Ross, M.L. Crisis of confidence averted: Impairment of exercise economy and performance in elite race walkers by ketogenic low carbohydrate, high fat (LCHF) diet is reproducible. *PLoS ONE* **2020**, *15*, e0234027. [CrossRef]
8. Cox, G.R.; Clark, S.A.; Cox, A.J.; Halson, S.L.; Hargreaves, M.; Hawley, J.A.; Burke, L.M. Daily training with high carbohydrate availability increases exogenous carbohydrate oxidation during endurance cycling. *J. Appl. Physiol.* **2010**, *109*, 126–134. [CrossRef] [PubMed]

9. Stellingwerff, T.; Spriet, L.L.; Watt, M.J.; Kimber, N.E.; Hargreaves, M.; Hawley, J.A.; Burke, L.M. Decreased PDH activation and glycogenolysis during exercise following fat adaptation with carbohydrate restoration. *Am. J. Physiol. Endocrinol. Metab.* **2006**, *290*, E380–E388. [CrossRef] [PubMed]

10. Hulston, C.J.; Venables, M.C.; Mann, C.H.; Martin, C.; Philp, A.; Baar, K.; Jeukendrup, A.E. Training with low muscle glycogen enhances fat metabolism in well-trained cyclists. *Med. Sci. Sports Exerc.* **2010**, *42*, 2046–2055. [CrossRef] [PubMed]

11. Yeo, W.K.; Paton, C.D.; Garnham, A.P.; Burke, L.M.; Carey, A.L.; Hawley, J.A. Skeletal muscle adaptation and performance responses to once a day versus twice every second day endurance training regimens. *J. Appl. Physiol.* **2008**, *105*, 1462–1470. [CrossRef] [PubMed]

12. Gleeson, M. Can nutrition limit exercise-induced immunodepression? *Nutr. Rev.* **2006**, *64*, 119–131. [CrossRef] [PubMed]

13. Gleeson, M. Mucosal immune responses and risk of respiratory illness in elite athletes. *Exerc. Immunol. Rev.* **2000**, *6*, 5–42. [PubMed]

14. Costa, R.J.; Jones, G.E.; Lamb, K.L.; Coleman, R.; Williams, J.H. The effects of a high carbohydrate diet on cortisol and salivary immunoglobulin A (s-IgA) during a period of increase exercise workload amongst Olympic and Ironman triathletes. *Int. J. Sports Med.* **2005**, *26*, 880–885. [CrossRef]

15. McKay, A.K.A.; Pyne, D.B.; Peeling, P.; Sharma, A.P.; Ross, M.L.R.; Burke, L.M. The impact of chronic carbohydrate manipulation on mucosal immunity in elite endurance athletes. *J. Sports Sci.* **2019**, *37*, 553–559. [CrossRef]

16. Shaw, D.M.; Merien, F.; Braakhuis, A.; Keaney, L.; Dulson, D.K. Adaptation to a ketogenic diet modulates adaptive and mucosal immune markers in trained male endurance athletes. *Scand. J. Med. Sci. Sports* **2020**, *31*, 140–152. [CrossRef]

17. Goedecke, J.H.; Christie, C.; Wilson, G.; Dennis, S.C.; Noakes, T.D.; Hopkins, W.G.; Lambert, E.V. Metabolic adaptations to a high-fat diet in endurance cyclists. *Metabolism* **1999**, *48*, 1509–1517. [CrossRef]

18. Burke, L.M.; Whitfield, J.; Heikura, I.A.; Ross, M.L.R.; Tee, N.; Forbes, S.F.; Hall, R.; McKay, A.K.A.; Wallett, A.M.; Sharma, A.P. Adaptation to a low carbohydrate high fat diet is rapid but impairs endurance exercise metabolism and performance despite enhanced glycogen availability. *J. Physiol.* **2020**. [CrossRef]

19. Schoffelen, P.; den Hoed, M.; van Breda, E.; Plasqui, G. Test-retest variability of VO_{2max} using total-capture indirect calorimetry reveals linear relationship of VO_2 and Power. *Scand. J. Med. Sci. Sports* **2019**, *29*, 213–222. [CrossRef]

20. Ackland, T.R.; Lohman, T.G.; Sundgot-Borgen, J.; Maughan, R.J.; Meyer, N.L.; Stewart, A.D.; Müller, W. Current status of body composition assessment in sport: Review and position statement on behalf of the ad hoc research working group on body composition health and performance, under the auspices of the I.O.C. Medical Commission. *Sports Med.* **2012**, *42*, 227–249. [CrossRef]

21. Meijboom, S.; van Houts-Streppel, M.T.; Perenboom, C.; Siebelink, E.; van de Wiel, A.M.; Geelen, A.; de Vries, J.H.M. Evaluation of dietary intake assessed by the Dutch self-administered web-based dietary 24-h recall tool (Compl-eat) against interviewer-administered telephone-based 24-h recalls. *J. Nutr. Sci.* **2017**, *6*, e49. [CrossRef] [PubMed]

22. Wendel-Vos, G.C.; Schuit, A.J.; Saris, W.H.; Kromhout, D. Reproducibility and relative validity of the short questionnaire to assess health-enhancing physical activity. *J. Clin. Epidemiol.* **2003**, *56*, 1163–1169. [CrossRef]

23. Salimetrics, S. Saliva Collection and Handling Advice. Available online: https://www.salimetrics.com/assets/documents/Saliva_Collection_Handbook.pdf (accessed on 20 May 2018).

24. Wilms, E.; Jonkers, D.; Savelkoul, H.F.J.; Elizalde, M.; Tischmann, L.; de Vos, P.; Troost, F.J. The Impact of Pectin Supplementation on Intestinal Barrier Function in Healthy Young Adults and Healthy Elderly. *Nutrients* **2019**, *11*, 1554. [CrossRef] [PubMed]

25. Barrett, B.; Brown, R.L.; Mundt, M.P.; Thomas, G.R.; Barlow, S.K.; Highstrom, A.D.; Bahrainian, M. Validation of a short form Wisconsin Upper Respiratory Symptom Survey (WURSS-21). *Health Qual. Life Outcomes* **2009**, *7*, 76. [CrossRef] [PubMed]

26. Volek, J.S.; Freidenreich, D.J.; Saenz, C.; Kunces, L.J.; Creighton, B.C.; Bartley, J.M.; Phinney, S.D. Metabolic characteristics of keto-adapted ultra-endurance runners. *Metabolism* **2016**, *65*, 100–110. [CrossRef]

27. Klein, S.; Coyle, E.F.; Wolfe, R.R. Fat metabolism during low-intensity exercise in endurance-trained and untrained men. *Am. J. Physiol.* **1994**, *267*, E934–E940. [CrossRef]

28. Hall, K.D.; Chen, K.Y.; Guo, J.; Lam, Y.Y.; Leibel, R.L.; Mayer, L.E.; Ravussin, E. Energy expenditure and body composition changes after an isocaloric ketogenic diet in overweight and obese men. *Am. J. Clin. Nutr.* **2016**, *104*, 324–333. [CrossRef]

29. Yeo, W.K.; Carey, A.L.; Burke, L.; Spriet, L.L.; Hawley, J.A. Fat adaptation in well-trained athletes: Effects on cell metabolism. *Appl. Physiol. Nutr. Metab.* **2011**, *36*, 12–22. [CrossRef]

30. Frayn, K.N. Calculation of substrate oxidation rates in vivo from gaseous exchange. *J. Appl. Physiol. Respir. Environ. Exerc. Physiol.* **1983**, *55*, 628–634. [CrossRef]

31. Mielgo Ayuso, J.; Zourdos, M.C.; Urdampilleta, A.; Calleja González, J.; Seco, J.; Córdova, A. Relationship of long-term macronutrients intake on anabolic-catabolic hormones in female elite volleyball players. *Nutr. Hosp.* **2017**, *34*, 1155–1162. [CrossRef]

32. Bergström, J.; Hermansen, L.; Saltin, B. Diet, muscle glycogen and physical performance. *Acta Phys. Scand.* **1967**, *71*, 140–150. [CrossRef] [PubMed]

33. Pitsiladis, Y.P.; Maughan, R.J. The effects of exercise and diet manipulation on the capacity to perform prolonged exercise in the heat and in the cold in trained humans. *J. Physiol.* **1999**, *517 Pt 3*, 919–930. [CrossRef]

34. Starling, R.D.; Trappe, T.A.; Parcell, A.C.; Kerr, C.G.; Fink, W.J.; Costill, D.L. Effects of diet on muscle triglyceride and endurance performance. *J. Appl. Physiol.* **1997**, *82*, 1185–1189. [CrossRef] [PubMed]

35. Bishop, N.C.; Walsh, N.P.; Haines, D.L.; Richards, E.E.; Gleeson, M. Pre-exercise carbohydrate status and immune responses to prolonged cycling: II. Effect on plasma cytokine concentration. *Int. J. Sport Nutr. Exerc. Metab.* **2001**, *11*, 503–512. [CrossRef] [PubMed]

36. Gleeson, M.; Blannin, A.K.; Walsh, N.P.; Bishop, N.C.; Clark, A.M. Effect of low- and high-carbohydrate diets on the plasma glutamine and circulating leukocyte responses to exercise. *Int. J. Sport Nutr.* **1998**, *8*, 49–59. [CrossRef]

37. Mortatti, A.L.; Moreira, A.; Aoki, M.S.; Crewther, B.T.; Castagna, C.; de Arruda, A.F.; Filho, J.M. Effect of competition on salivary cortisol, immunoglobulin A, and upper respiratory tract infections in elite young soccer players. *J. Strength Cond. Res.* **2012**, *26*, 1396–1401. [CrossRef]

38. Campbell, J.P.; Turner, J.E. Debunking the Myth of Exercise-Induced Immune Suppression: Redefining the Impact of Exercise on Immunological Health Across the Lifespan. *Front. Immunol.* **2018**, *9*, 648. [CrossRef]

39. Davison, G.; Kehaya, C.; Wyn Jones, A. Nutritional and Physical Activity Interventions to Improve Immunity. *Am. J. Lifestyle Med.* **2016**, *10*, 152–169. [CrossRef]

40. Gunzer, W.; Konrad, M.; Pail, E. Exercise-induced immunodepression in endurance athletes and nutritional intervention with carbohydrate, protein and fat-what is possible, what is not? *Nutrients* **2012**, *4*, 1187–1212. [CrossRef]

41. Vargas, S.; Romance, R.; Petro, J.L.; Bonilla, D.A.; Galancho, I.; Espinar, S.; Benitez-Porres, J. Efficacy of ketogenic diet on body composition during resistance training in trained men: A randomized controlled trial. *J. Int. Soc. Sports Nutr.* **2018**, *15*, 31. [CrossRef]

42. Volek, J.S.; Sharman, M.J.; Love, D.M.; Avery, N.G.; Gomez, A.L.; Scheett, T.P.; Kraemer, W.J. Body composition and hormonal responses to a carbohydrate-restricted diet. *Metabolism* **2002**, *51*, 864–870. [CrossRef]

43. Kreitzman, S.N.; Coxon, A.Y.; Szaz, K.F. Glycogen storage: Illusions of easy weight loss, excessive weight regain, and distortions in estimates of body composition. *Am. J. Clin. Nutr.* **1992**, *56* (Suppl. 1), 292s–293s. [CrossRef]

44. Bingham, S.A. Limitations of the various methods for collecting dietary intake data. *Ann. Nutr. Metab.* **1991**, *35*, 117–127. [CrossRef]

45. Mielgo-Ayuso, J.; Calleja-González, J.; Refoyo, I.; León-Guereño, P.; Cordova, A.; Del Coso, J. Exercise-Induced Muscle Damage and Cardiac Stress During a Marathon Could be Associated with Dietary Intake During the Week Before the Race. *Nutrients* **2020**, *12*, 316. [CrossRef]

Article

Additive Effects of Omega-3 Fatty Acids and Thiazolidinediones in Mice Fed a High-Fat Diet: Triacylglycerol/Fatty Acid Cycling in Adipose Tissue

Kristina Bardova [1,†], Jiri Funda [1,†], Radek Pohl [2], Tomas Cajka [3,4], Michal Hensler [1], Ondrej Kuda [5], Petra Janovska [1], Katerina Adamcova [1], Ilaria Irodenko [1], Lucie Lenkova [1], Petr Zouhar [1], Olga Horakova [1], Pavel Flachs [1,‡], Martin Rossmeisl [1], Jerry Colca [6] and Jan Kopecky [1,*]

[1]   Laboratory of Adipose Tissue Biology, Institute of Physiology of the Czech Academy of Sciences, Videnska 1083, 142 20 Prague 4, Czech Republic; kristina.bardova@fgu.cas.cz (K.B.); jiri.funda@fgu.cas.cz (J.F.); hensler.michal@gmail.com (M.H.); petra.janovska@fgu.cas.cz (P.J.); katerina.adamcova@fgu.cas.cz (K.A.); ilaria.irodenko@fgu.cas.cz (I.I.); lucie.lenkova@fgu.cas.cz (L.L.); petr.zouhar@fgu.cas.cz (P.Z.); olga.horakova@fgu.cas.cz (O.H.); flachs@biomed.cas.cz (P.F.); martin.rossmeisl@fgu.cas.cz (M.R.)

[2]   NMR Spectroscopy, Institute of Organic Chemistry and Biochemistry of the Czech Academy of Sciences, Flemmingovo Namesti 542/2, 160 00 Prague 6, Czech Republic; radek.pohl@uochb.cas.cz

[3]   Laboratory of Metabolomics, Institute of Physiology of the Czech Academy of Sciences, Videnska 1083, 142 20 Prague 4, Czech Republic; tomas.cajka@fgu.cas.cz

[4]   Laboratory of Translational Metabolism, Institute of Physiology of the Czech Academy of Sciences, Videnska 1083, 142 20 Prague 4, Czech Republic

[5]   Laboratory of Metabolism of Bioactive Lipids, Institute of Physiology of the Czech Academy of Sciences, Videnska 1083, 142 20 Prague 4, Czech Republic; ondrej.kuda@fgu.cas.cz

[6]   Cirius Therapeutics, Kalamazoo, MI 490 07, USA; jcolca@ciriustx.com

*   Correspondence: jan.kopecky@fgu.cas.cz; Tel.: +420-296442554; Fax: +420-296442599

†   These authors contributed equally to this work.

‡   Deceased.

Received: 6 November 2020; Accepted: 2 December 2020; Published: 4 December 2020

Abstract: Long-chain n-3 polyunsaturated fatty acids (Omega-3) and anti-diabetic drugs thiazolidinediones (TZDs) exhibit additive effects in counteraction of dietary obesity and associated metabolic dysfunctions in mice. The underlying mechanisms need to be clarified. Here, we aimed to learn whether the futile cycle based on the hydrolysis of triacylglycerol and re-esterification of fatty acids (TAG/FA cycling) in white adipose tissue (WAT) could be involved. We compared Omega-3 (30 mg/g diet) and two different TZDs—pioglitazone (50 mg/g diet) and a second-generation TZD, MSDC-0602K (330 mg/g diet)—regarding their effects in C57BL/6N mice fed an obesogenic high-fat (HF) diet for 8 weeks. The diet was supplemented or not by the tested compound alone or with the two TZDs combined individually with Omega-3. Activity of TAG/FA cycle in WAT was suppressed by the obesogenic HF diet. Additive effects in partial rescue of TAG/FA cycling in WAT were observed with both combined interventions, with a stronger effect of Omega-3 and MSDC-0602K. Our results (i) supported the role of TAG/FA cycling in WAT in the beneficial additive effects of Omega-3 and TZDs on metabolism of diet-induced obese mice, and (ii) showed differential modulation of WAT gene expression and metabolism by the two TZDs, depending also on Omega-3.

Keywords: insulin; lipogenesis; obesity; glucose homeostasis; adipocytes

1. Introduction

The prevention and treatment of obesity and associated pathologies, namely, dyslipidemia, type 2 diabetes, hepatosteatosis, cardiovascular disease and cancer, represent major challenges for the healthcare system. Reflecting on the multiple pathological mechanisms involved, any treatment must include, besides pharmacological interventions, lifestyle changes, including dietary measures. Long-chain n-3 polyunsaturated fatty acids (Omega-3) in the diet, such as eicosapentaenoic acid (EPA; 20:5n-3) and docosahexaenoic acid (DHA; 22:6n-3), exert broad benefits on health (reviewed in [1–5]). These lipids act as natural hypolipidemics, reduce accumulation of hepatic lipids ([6,7]; reviewed in [8]), increase plasma levels of adiponectin ([9]; reviewed in [10]), ameliorate low-grade inflammation in obesity [11] and enhance intestinal fatty acid (FA) oxidation [12,13]. Moreover, Omega-3s reduce adiposity while limiting proliferation of adipocytes in rodent models of obesity [14,15] and could reduce obesity, even in humans [16,17]. Animal experiments document the beneficial effects of Omega-3 on insulin sensitivity and glucose metabolism even under conditions of established insulin resistance. Although Omega-3s cannot reverse insulin resistance in diabetic patients, some intervention studies document their positive effects on glucose homeostasis in prediabetic subjects (reviewed in [18]).

In diet-induced obese mice, we have observed additive effects of Omega-3 and anti-diabetic drugs thiazolidinediones (TZDs) in counteracting obesity and alleviating dyslipidemia, hepatosteatosis, low-grade inflammation of white adipose tissue (WAT) and whole-body insulin resistance. Induction of adiponectin and improvement of insulin sensitivity in muscle were also observed, in the absence of any effect on food consumption [19,20]. Several TZDs were used to improve insulin sensitivity in patients with type 2 diabetes, resulting from the interaction of TZDs with (i) peroxisome-proliferator activated receptor γ (PPARγ), the transcription factor representing the dominant regulator of adipogenesis and fat cell gene expression [21]; and (ii) mitochondrial pyruvate carrier, which is inhibited by TZD binding [22,23]. However, the "classical" TZDs induce side effects that are mediated by PPARγ. Thus, pioglitazone is the only TZD remaining in clinical use [24]. Second-generation, "PPARγ-sparing" TZDs were developed to reduce the adverse side-effects of TZDs. These novel TZDs minimize direct binding to PPARγ [25] but retain the ability to decrease activity of the mitochondrial pyruvate carrier [25–27]. Thus, in s recent phase 2b clinical trial, MSDC-0602K decreased fasting glucose, insulin, glycated hemoglobin and markers of liver injury without dose-limiting side effects [28].

Adverse consequences of obesity are reflected in large ectopic accumulations of lipids in extra-adipose tissues. This "lipotoxic" impact depends in part on the insufficient capacity of WAT to buffer plasma non-esterified fatty acid (NEFA) levels. Thus, intrinsic metabolic properties of WAT play a role, namely, in the activity of the futile cycle based on the hydrolysis of triacylglycerols (TAG) and re-esterification of FAs in adipocytes (TAG/FA cycling). This core biochemical activity of WAT is required for fine and fast tuning of plasma NEFA levels [1,29–32]. Together with in situ FA synthesis (de novo lipogenesis; DNL), TAG/FA cycling is linked to oxidative phosphorylation (OXPHOS), because ATP is required for both of these processes [1,2,33]. Moreover, TAG synthesis requires constant generation of glycerol 3-phosphate. In WAT, its formation occurs mainly via glycolysis and also from precursors other than glucose, i.e., via glyceroneogenesis, whereas direct phosphorylation of glycerol is of a minor significance [29,34,35].

The activity of TAG/FA cycling in WAT is under complex control, as inferred by its positioning at the intersection of various metabolic fluxes in adipocytes (reviewed in [1,2,33]). Our results indicated that beneficial effects of Omega-3 on plasma lipids, glucose homeostasis and liver fat accumulation in diet-induced obese mice were linked to the stimulation of mitochondrial biogenesis [36] and OXPHOS activity [6] in epididymal WAT (eWAT), and probably also TAG/FA cycling in this tissue [37]. TZDs induce TAG/FA cycling in WAT via stimulation of glyceroneogenesis [29,38]. Whether WAT metabolism can be involved in the beneficial systemic effects of the combined intervention using Omega-3 and TZDs has not yet been studied [19,20,39,40].

Therefore, we have tested the effect of the combined intervention using the model of obesity induced by a high-fat (HF) diet in mice. The experiments were conducted similarly as before [19,20,39], except that in addition to the "classical" TZD pioglitazone, a second-generation TZD, namely, MSDC-0602K, was also used. Moreover, the activity of TAG/FA cycling in eWAT was evaluated in vivo, in parallel with the characterization of eWAT gene expression and other selected whole-body phenotypes. Our results support the role of TAG/FA cycling in WAT for the beneficial additive effects of Omega-3s and TZDs on the metabolism of diet-induced obese mice. We showed differential modulation of eWAT gene expression and metabolism by the two TZDs, depending also on Omega-3.

2. Materials and Methods

2.1. Animals and Dietary Interventions

Experiments were performed similarly as before [13,19,20,39]. Thus, male C57BL/6N mice (Charles River Laboratories, Sulzfeld, Germany) were maintained at 22 °C in a 12 h light–dark cycle (light from 6.00 a.m.) with free access to water and standard chow diet (STD; 3.4% wt/wt as lipids; rat/mouse—maintenance extrudate; Ssniff Spezialdieten GmbH, Soest, Germany). At 3 months of age, mice were randomly assigned ($n = 8$–10) to the HF diet (lipid content, ≈35% wt/wt, mainly corn oil; [19]) or to the following "interventions," which were based on feeding (i) HF + F, a HF diet supplemented with Omega-3 concentrate (46% DHA, 14% EPA, wt/wt, as TAG; product EPAX 1050 TG; EPAX a.s., Lysaker, Norway), which replaced 15% wt/wt of dietary lipids to achieve a total EPA and DHA concentration of 30 mg/g diet (tocopherol content 0.02% wt/wt); (ii) HF + PIO, a HF diet supplemented with 50 mg pioglitazone/kg diet (Actos; Takeda, Japan); (iii) HF + PIO + F, a HF diet supplemented with both pioglitazone and Omega-3; (iv) HF + MSDC, a HF diet supplemented with 330 mg MSDC-0602K/kg diet (Cirius Therapeutics, USA); and (v) HF + MSDC + F, a HF diet supplemented with both MSDC-0602K and Omega-3. The dose for supplementation of HF diet by pioglitazone and Omega-3 was as before, i.e., under the conditions when the additive effects of the two interventions could be observed [20]. The dose of MSDC-0602K was the same as in the previous studies using this TZD in mice [25,26,41]. Diets were stored at −20 °C, in sealed plastic bags filled with nitrogen. The animals received fresh aliquots of the diet every other day. For the composition of macronutrients and the FA profiles in HF diets (see ESM Table 2 of ref. [19], where cHF and cHF + F represent HF and HF + F, respectively, of the present study). Some mice were maintained on the STD to serve as lean controls. Mice were fed the respective diets for 8 weeks. Body weights were recorded every other week, while 24 h food consumption was measured every week.

Mice were sacrificed in a non-fasted state by cervical dislocation under diethyl ether anesthesia (between 8.00 and 10.00 a.m.). Liver, eWAT and subcutaneous WAT in the inguinal and gluteal region (scWAT) were dissected and snap frozen in liquid nitrogen; ethylenediaminetetraacetic acid (EDTA) treated plasma was collected; and all the samples were stored at −80 °C. The experiments followed the guidelines for the use and care of laboratory animals of the Institute of Physiology of the Czech Academy of Sciences and were approved under protocol number 81/2016.

2.2. Biochemical Analysis of Plasma and Tissue Samples

Plasma levels of (i) TAG and total cholesterol were determined using the colorimetric enzymatic assays from Erba Lachema (Brno, Czech Republic), and (ii) NEFAs were assessed with a NEFA-HR(2) kit from Waco Chemicals GmbH (Neuss, Germany). Blood glucose levels were measured by OneTouch Ultra glucometers (LifeScan, Milpitas, CA, USA), and plasma insulin levels were determined by the Sensitive Rat Insulin RIA Kit (Millipore, Billerica, MA, USA). Liver TAG content was estimated in ethanolic KOH tissue solubilisates as before [13].

2.3. Glucose Homeostasis

Intraperitoneal glucose tolerance test (GTT) was performed using 1 mg of glucose/g body weight in overnight fasted mice. The homeostatic model assessment of insulin resistance (HOMA) index was calculated as described [20].

2.4. Histological and Immunohistological Analysis of eWAT

As described before [6], formalin-fixed paraffin-embedded tissue sections stained by hematoxylin and eosin were used for adipocyte morphometry (600 objects per section were evaluated). A macrophage marker MAC-2/galectin-3 was detected using specific antibodies to calculate a relative density of crown-like structures (CLS). Digital images were captured using Olympus AX70 light microscope and a DP 70 camera (Olympus, Tokyo, Japan). Images were analyzed using NIS Elements software (Laboratory Imaging, Prague, Czech Republic). All histological analyses were performed by a pathologist blinded to dietary groups.

2.5. Real-time Quantitative PCR (RT-qPCR)

Total tissue RNA was isolated and gene expression was evaluated as described [15]. Data were normalized [42] to the geometric mean signal of four reference genes (*Eef2*, *Eef1a1*, *Actb* and *Cyphb/Ppib*). Background gene expression levels were defined by the mean Cp value (the cycle at which the fluorescence of a sample rises above the background fluorescence) lower than 30.

Gene abbreviations and an overview of primer sequences are given in Table 1.

Table 1. Sequences of primers.

Gene Name	Gene ID	Forward Primer	Reverse Primer
Acsl1	14081	GAAGCCGTGGCCCAGGTGTTTGTC	TTCGCCTTCAGTGTTGGAGTCAGA
Actb	11461	GAACCCTAAGGCCAACCGTGAAAAGAT	ACCGCTCGTTGCCAATAGTGATG
Atgl/Pnpla2	66853	GGCAATCAGCAGGCAGGGTCTTTA	GCCAACGCCACTCACATCTACG
Bckdha	12039	ACGGCGGGCTGTGGCTGAGAA	GAGATTGGGTGGTCCTGCTTGTCC
Cd36	12491	TGATACTATGCCCGCCTCTCC	TTCCCACACTCCTTTCTCCTCTAC
Cpt1a	12894	GCAGCTCGCACATTACAAGGACAT	AGCCCCCGCCACAGGACACATAGT
Crat	12908	ACATGGTGGTGGTAGCAAGTTCAA	GGCAAGGGCACCATAGGAGA
Cyphb/Ppib	19035	GGGAGATGGCACAGGAGGAAAGAG	ACCCAGCCAGGCCCGTAGTG
Dgat1	13350	TGGCCAGGACAGGAGTATTTTTGA	CTCGGGCATCGTAGTTGAGCA
Dgat2	67800	TGCCCTACTCCAAGCCCATCACC	TCAGTTCACCTCCAGCACCTCAGTCTC
Eef1a1	13627	TGACAGCAAAAACGACCCACCAAT	GGGCCATCTTCCAGCTTCTTACCA
Eef2	13629	GAAACGCGCAGATGTCCAAAAGTC	GCCGGGCTGCAAGTCTAAGG
Elovl5	68801	CCTCTCGGGTGGCTGTTCTTCC	AGGCTTCGGCTCGGCTTGTC
Fas/Fasn	14104	TGGGTGTGGAAGTTCGTCAG	GTCGTGTCAGTAGCCGAGTC
Gk	14933	TCGTTCCAGCATTTTCAGGGTTAT	TCAGGCATGGAGGGTTTCACTACT
Hsl/Lipe	16890	TGCGCCCCACGGAGTCTATGC	CTCGGGGCTGTCTGAAGGCTCTGA
Lpl	16956	AGCCCCCAGTCGCCTTTCTCCT	TGCTTTGCTGGGGTTTTCTTCATTCA
Mpc1	55951	TCATTCAGGGAGGACGACTTATC	TGTTTTCCCTTCAGCACGACTAC
Mpc2	70456	CTCCCACCCTGCTGCTGTCG	GGCCTGCCGGGTGGTTGTA
Pc/Pcx	18563	CCCCTGGATAGCCTTAATACTCGT	TGGCCCTTCACATCCTTCAAA
Pck1	18534	GGCAGCATGGGGTGTTTGTAGGA	TTTGCCGAAGTTGTAGCCGAAGAAG
Pdk4	27273	GGCTTGCCAATTTCTCGTCTCTA	TTCGCCAGGTTCTTCGGTTCC
Pgc1a/Ppargc1a	19017	CCCAAAGGATGCGCTCTCGTT	TGCGGTGTCTGTAGTGGCTTGATT
Pparg	19016	GCCTTGCTGTGGGGATGTCTC	CTCGCCTTGGCTTTGGTCAG
Sbacad/Acadsb	66885	GCATCTGAGGTCGCTGGGCTAAC	CGATGTGCTTGGCGATGGTGT
Ucp1	22227	CACGGGGACCTACAATGCTTACAG	CACGGGGACCTACAATGCTTACAG
Vlcad/Acadvl	11370	CAGGGGTGGAGCGTGTGC	CATTGCCCAGCCCAGTGAGTTCC

Abbreviations: *Acsl1*, acyl-CoA synthetase long-chain family member 1 (ACSL1); *Actb*, actin beta (ACTB); *Atgl/Pnpla2*, adipose triglyceride lipase, also known as patatin-like phospholipase domain-containing protein 2 (ATGL); *Bckdha*, branched chain ketoacid dehydrogenase E1, alpha polypeptide (BCKDHA); *Cpt1a*, carnitine palmitoyltransferase 1a (CPT1A); *Crat*, carnitine acetyltransferase (CRAT); *Cyphb/Ppib*, cyclophilin beta, also known as peptidylprolyl isomerase B (CYPHB); *Dgat1*, diacylglycerol O-acyltransferase 1 (DGAT1); *Dgat2*, diacylglycerol O-acyltransferase 2 (DGAT2); *Eef1a1*, eukaryotic translation elongation factor 1 alpha (EEF1A1); *Eef2*, eukaryotic translation elongation factor 2 (EEF2); *Elovl5*, ELOVL family member 5, elongation of long-chain fatty acids (ELOVL5); *Fas/Fasn*, fatty acid synthase (FAS); *Gk*, glycerol kinase (GK); *Hsl/Lipe*, lipase, hormone sensitive (HSL); *Lpl*, lipoprotein lipase (LPL); *Mpc1*, mitochondrial pyruvate carrier 1 (MPC1); *Mpc2*, mitochondrial pyruvate carrier 2 (MPC2); *Pc/Pcx*, pyruvate carboxylase (PC); *Pck1*, phosphoenolpyruvate carboxykinase 1, cytosolic (PCK1); *Pdk4*, pyruvate dehydrogenase kinase, isoenzyme 4 (PDK4); *Pgc1a/ Ppargc1a*, peroxisome proliferative activated receptor gamma, coactivator 1 alpha (PGC1A); *Pparg* peroxisome proliferator activated receptor gamma (PPARG); *Sbacad/Acadsb*, acyl-Coenzyme A dehydrogenase, short/branched chain (SBACAD); *Ucp1*, uncoupling protein 1 (UCP1); *Vlcad/Acadvl*, acyl-Coenzyme A dehydrogenase, very long chain (VLCAD).

2.6. In Vivo Evaluation of TAG synthesis and DNL-derived FA in eWAT

TAG synthesis and DNL were characterized using in vivo ^{2}H enrichment of TAG similarly to before [33]. Due to the suppression of TAG/FA cycling activity by HF diet (see Results), the period of in vivo ^{2}H-incorporation into TAG was extended to 21 days before dissection, whereas the ^{2}H$_2$O concentration in drinking water was increased to 10% and the intraperitoneal ^{2}H$_2$O bolus was omitted. This labeling was performed in 8 out of 10 mice per group; 2 mice served as a negative control for ^{2}H$_2$O analyses. After dissection, lipids from eWAT were extracted similarly as for FAHFA extraction in [43], except that the samples were homogenized in a mixture of citric acid and ethylacetate. Dried organic phase was resuspended in hexane and applied on Discovery DSC-Si SPE tubes (52 μm, 72 Å; MERCK, Darmstadt, Germany). TAG fraction was eluted from SPE tubes with a mixture of hexane and MTBE. Samples were analyzed using either nuclear magnetic resonance (NMR) spectroscopy or liquid chromatography–mass spectrometry (LC–MS).

^{1}H and ^{2}H-NMR spectroscopy was performed as before [33] using AVANCE III HD 500 MHz system (Bruker Corporation) equipped with ^{19}F lock and a 5-mm CP BBO-^{1}H&^{19}F-^{2}H probe. The spectra were analyzed using MestReNova and spectral deconvolution was used in case of ^{2}H for integration of signals. The amounts of ^{1}H and ^{2}H in both glycerol and fatty-acyl moieties of TAG were calculated from the peak area relative to the peak of the pyrazine ^{1}H/^{2}H standard. Since (i) ^{2}H can be incorporated in a glycerol moiety of TAG only before esterification of FA to glycerol, and (ii) glycerol formed during lipolysis in WAT is assumed to be released into the circulation and not converted to glycerol-3-phosphate in situ (reviewed in [1]; see Discussion), TAG positional ^{2}H enrichment of the glycerol moiety reflects the rate of TAG synthesis. Enrichment of newly synthesized glycerol-3-phosphate from ^{2}H$_2$O was assumed to be stoichiometric for all five positional hydrogens regardless of the relative contributions of glycolysis and glyceroneogenesis [34]; however, relative contributions of the individual carbons to the labeling of the glycerol moiety of TAG might differ. Therefore, the ^{2}H enrichment of the glycerol moiety was evaluated separately for (i) sn 1 + 3 and (ii) sn 2 carbon. Similarly, ^{2}H enrichment of FA methyls in TAG correlates with de novo FA synthesis (DNL) rate. Measurement of fractional TAG/FA cycling drew on previously validated assumptions of glycerol ^{2}H-enrichment from body ^{2}H-enriched water [34].

Analysis of both TAG species and their FA after hydrolysis was conducted using LC–MS. Details can be found in Supplemental Materials. In total 135 deuterated TAG species were detected; 62 species above basal were considered, representing 98% of lipids. In total, 44 deuterated FA above basal were detected in TAG hydrolysates; however, only 8 of them, which were detected in all the groups, were considered (Supplementary Table S1). In order to characterize FA desaturation index, the product/substrate ratios (FA 14:1/14:0, FA 16:1/16:0 and FA 18:1/18:0) were calculated from peak heights obtained from LC–MS analysis.

2.7. Statistical Analysis

Data are presented as means ± SEM. Statistical analysis was performed using GraphPad Prism (Version 8.3.1, 2019). Dallal and Wilkinson's approximation to Lilliefors' method was used to test normality of the data and $\log_{10}$ transformed when needed. Data were analyzed in two ways: (i) to compare STD with individual HF-based diets, a two-tailed Student's t-test or Mann–Whitney's non-parametric test was used; and (ii) to compare HF-based diets with each other (i.e., to reveal the effects of various interventions), one-way ANOVA followed by the Tukey's post-hoc test or the Kruskal–Wallis non-parametric test with Dunn's post-hoc test were done. A $p \leq 0.05$ was considered to be significant. Partial least squares-discriminant analysis (PLS-DA) was performed using MetaboAnalyst 4.0 [44]. Significant outliers (according to Grubb's test) were excluded from further analyses.

3. Results

3.1. Parameters of Energy Balance, Adiposity and Lipid Metabolism Markers

Body weight was increased by HF diet administration. This obesogenic effect was reduced by all the interventions except HF + PIO and HF + F, which only tended to reduce body weight. The differential effects of diets on body weight became apparent already after 2 weeks of feeding (Figure 1). At the end of the 8-week-study, almost all the interventions significantly prevented body weight gain vs. HF diet (Figure 1 and Table 2): strongest effects for HF + MSDC vs. HF + F and HF + PIO. The combination interventions exerted even stronger effects than individual interventions—the highest ($\approx$2.7-fold) suppression of body weight gain was seen for the HF + MSDC + F mice. None of the HF-based diets affected food consumption (Table 2).

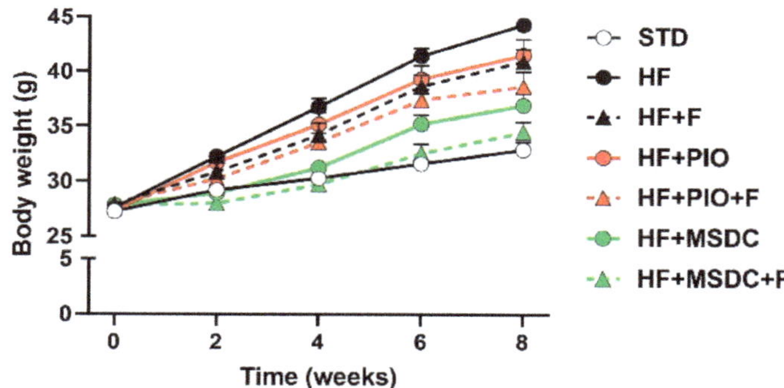

Figure 1. Body weights during dietary interventions. At 3 months of age, subgroups of mice (n = 8–10) were fed STD or various high fat (HF) diets for 8 weeks. Data were pooled from two separate experiments (resulting in n = 16–19/group). Data are means ± SEMs. For statistical differences at week 8, see Table 2. For designation of dietary groups, see Section 2.1. STD: standard chow diet; HF: high-fat; PIO: pioglitazone.

Table 2. Body mass, tissue weights, tissue TAG content and plasma parameters in mice fed various diets.

	STD	HF	HF + F	HF + PIO	HF + PIO + F	HF + MSDC	HF + MSDC + F
Energy balance							
Body weight initial g)	27.2 ± 0.40	27.6 ± 0.40	27.9 ± 0.49	27.3 ± 0.47	27.9 ± 0.48	27.8 ± 0.42	27.8 ± 0.50
Body weight final (g)	32.9 ± 0.70	44.3 ± 0.64 *	41.0 ± 1.13 *	41.5 ± 1.42 *	38.6 ± 1.35 *,a	36.9 ± 1.16 *,a	34.5 ± 0.91 abc
Body weight gain (g)	5.50 ± 0.44	16.1 ± 0.63 *	12.0 ± 0.75 *,a	13.3 ± 1.17 *	10.1 ± 1.11 *,a	8.27 ± 0.94 abc	6.02 ± 0.58 abcd
Cumulative food intake (MJ/animal)	3.86 ± 0.06	4.59 ± 0.11 *	4.21 ± 0.18	4.57 ± 0.25 *	4.17 ± 0.15	4.75 ± 0.20 *	4.81 ± 0.32 *
Tissues							
eWAT weight (mg)	833 ± 73	2489 ± 84 *	1977 ± 88 *,a	2252 ± 168 *	1734 ± 142 *,ac	1511 ± 123 *,ac	1171 ± 117 *,abcd
scWAT weight (mg)	403 ± 34	1398 ± 87 *	1283 ± 137 *	1614 ± 288 *	1191 ± 84 *	1097 ± 96 *	936 ± 78 *
Liver weight (mg)	1763 ± 43	1793 ± 63	1563 ± 58 *,a	1613 ± 84 *	1425 ± 58 *,a	1477 ± 58 *,a	1418 ± 33 *,a
Liver TAG content (mg/g)	33.2 ± 1.16	93.4 ± 8.03 *	57.0 ± 4.96 *,a	112 ± 17.6 *,b	48.6 ± 5.77 *,ac	67.9 ± 10.0 *c	42.4 ± 5.58 ac
Plasma (random fed state)							
NEFA (mmol/L)	0.50 ± 0.05	0.66 ± 0.04 *	0.38 ± 0.03 *,a	0.42 ± 0.02 a	0.34 ± 0.03 *,a	0.38 ± 0.04 *,a	0.22 ± 0.02 *,abce
TAG (mmol/L)	1.39 ± 0.06	1.79 ± 0.05 *	1.05 ± 0.09 *,a	0.99 ± 0.06 *,a	0.82 ± 0.06 *,a	0.89 ± 0.06 *,a	0.54 ± 0.04 *,abce
Cholesterol (mmol/L)	2.81 ± 0.33	4.61 ± 0.14 *	3.40 ± 0.12 a	3.91 ± 0.10 *,ab	2.91 ± 0.13 ac	3.40 ± 0.13 ac	2.55 ± 0.07 abce

Data are means ± SEMs (n = 16–19). Cumulative energy intake was assessed during the initial 7-week period of dietary interventions. * Significantly different from STD ($p \leq 0.05$, t-test). [a] Significantly different vs. HF; [b] significantly different vs. HF + F; [c] significantly different vs. HF + PIO; [d] significantly different vs. HF + PIO + F; [e] significantly different vs. HF + MSDC ($p \leq 0.05$, one-way ANOVA). STD: standard chow diet; HF: high-fat; PIO: pioglitazone; eWAT: epididymal white adipose tissue; scWAT: subcutaneous white adipose tissue; TAG: triacylglycerol; NEFA, non-esterified fatty acid. For designation of dietary groups, see Section 2.1.

Suppression of the obesogenic effect of HF diet was mirrored by changes in weight of eWAT (elevated in response to HF diet), which was $\approx$2.1-fold lower in the HF + MSDC + F group compared to the HF group. Similar changes in adiposity were observed at the level of scWAT, but differences between the diets were less pronounced (Table 2). Histological and immunohistochemical analysis of eWAT revealed adipocyte hypertrophy induced by HF diet, which was significantly prevented only by HF + MSDC + F (Figure 2A,B). HF diet also induced low-grade eWAT inflammation, marked by CLS that are formed by macrophages aggregated around dying adipocytes [45]. This infiltration was completely counteracted by HF + MSDC + F and markedly reduced by all other interventions except HF + F (Figure 2C,D).

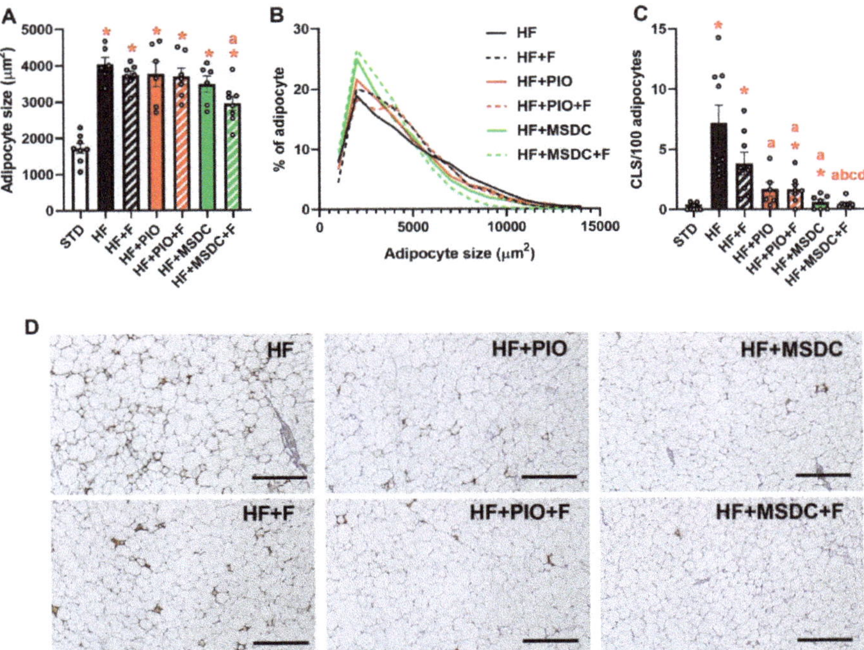

Figure 2. Morphology and immunohistochemistry of epididymal white adipose tissue (eWAT). Mice were fed STD or various HF diets for 8 weeks. Size of eWAT adipocytes was evaluated from hematoxylin + eosin stained sections (**A**) and expressed as mean (**A**) or histogram of adipocyte size ((**B**) pooled data from 4800 adipocytes). Inflammation of eWAT was assessed using the anti-Mac-2 antibody, which labeled macrophages aggregated around dying adipocytes ((**D**) reddish color; representative histological sections are shown; scale bar = 200 μm), and was expressed as the percentage of CLS/100 adipocytes (**C**). Data are means ± SEMs ($n = 8$). * Significantly different from STD ($p \leq 0.05$, t-test). a—significantly different vs. HF; b—significantly different vs. HF + F; c—significantly different vs. HF + PIO; d—significantly different vs. HF + PIO + F ($p \leq 0.05$, one-way ANOVA). For designation of dietary groups, see Section 2.1.

Liver weight was similar in the STD, HF and HF + PIO mice, but it was reduced ($\approx$1.1 to $\approx$1.3-fold) in response to all the other interventions. Liver TAG content was $\approx$2.8-fold higher in the HF mice compared to the STD mice. HF + PIO tended to increase liver TAG content vs. HF, whereas it was reduced by both HF + F and HF + MSDC compared to HF + PIO. It was also decreased by both combined interventions compared with the HF diet and HF + PIO. The effect of HF + MSDC + F tended to be the most pronounced ($\approx$2.2 and $\approx$1.6-fold decrease compared to the HF and HF + MSDC mice, respectively; Table 2).

3.2. Lipid and Glucose Homeostasis

Levels of plasma lipids, i.e., NEFA, TAG and cholesterol, were increased by HF diet compared to STD. All the interventions decreased plasma levels of these analytes, with the most pronounced effects being observed in the HF + MSDC + F mice. In these mice, the levels of all these lipid metabolites tended to be lower as compared with the HF + PIO + F mice (Table 2).

HF diet increased both fasting blood glucose and plasma insulin levels. These effects were completely counteracted by both combined interventions (Figure 3A,B). The remaining interventions showed similar, though less pronounced influences. Changes in HOMA index indicated a complete normalization of glucose homeostasis by HF + MSDC + F and significant improvements by the remaining interventions (Figure 3C). Furthermore, the deterioration of glucose clearance during intraperitoneal GTT observed in the HF mice was ameliorated by all the interventions (except for HF + F), which all exerted similar effects (Figure 3D,E).

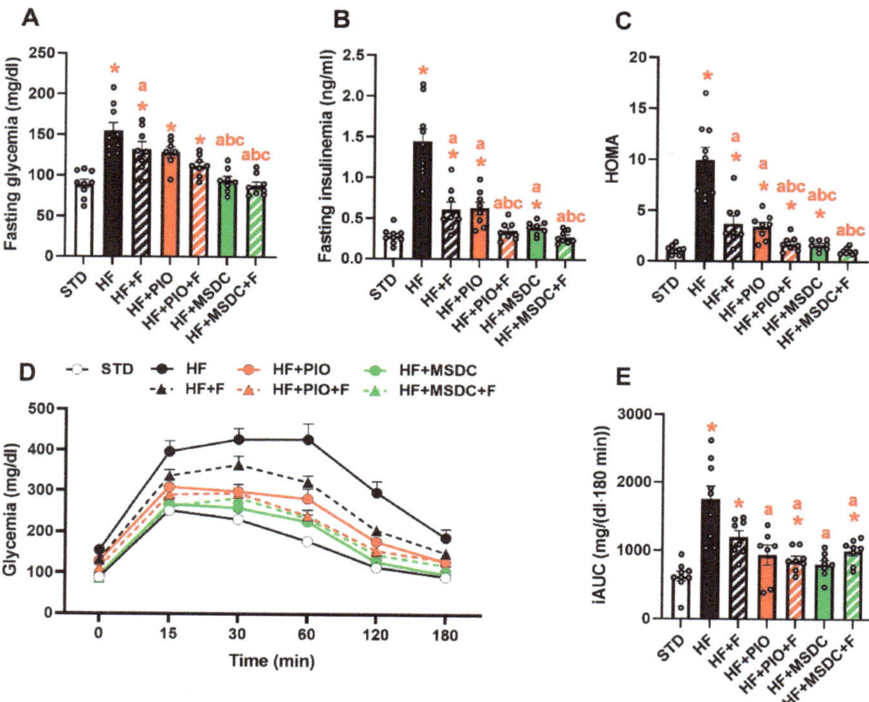

Figure 3. Glucose homeostasis. Mice were fed STD or various HF diets for 8 weeks. Glycemia (**A**) and insulinemia (**B**) after overnight fasting; HOMA index (**C**); intraperitoneal GTT: time course of glycemia during the test (**D**) and the corresponding incremental area under the curve (iAUC; (**E**)). Data are means ± SEMs ($n = 8$). * Significantly different from STD ($p \leq 0.05$, *t*-test). a—significantly different vs. HF; b—significantly different vs. HF + F; c—significantly different vs. HF + PIO ($p \leq 0.05$, one-way ANOVA). HOMA: homeostatic model assessment; GTT: glucose tolerance test. For designation of dietary groups, see Section 2.1.

3.3. In Vivo Evaluation of TAG/FA Cycling Activity in eWAT

Next, we focused on the main goal of this study, verification of the hypothesis that the combined interventions using TZDs and Omega-3 exert an additive effect on TAG/FA cycling in WAT of mice fed an obesogenic HF diet. Therefore, this biochemical activity was characterized in eWAT, a typical WAT depot, using ^2H_2O administration in vivo and subsequent analysis of ^2H enrichment of TAG.

Analysis of deuterated TAG using NMR spectroscopy enabled for separate evaluations of total enrichment of ^2H in glycerol and FA methyl moieties of TAG, reflecting the rates of TAG glycerol 3-phosphate precursor synthesis (Figure 4A,B) and DNL activity (Figure 4C), respectively. The ^2H enrichment of the glycerol moiety was evaluated separately for sn 1 + 3 (Figure 4A) and sn 2 positions (Figure 4B). The enrichment at all these positions was suppressed by HF diet compared with STD, ≈1.6 and ≈2.6-fold, respectively (Figure 4A,B). There was no significant effect of any intervention (i.e., difference between the HF-based diets) on sn 1 + 3 position if evaluated by 1-way ANOVA (Figure 4A), but there was a significant effect of TZDs if evaluated by 2-way ANOVA (one factor—Omega-3; another factor—TZD). The ^2H enrichment at sn 2 position was affected more strongly: HF + F, HF + PIO and HF + MSDC tended to increase the enrichment, while both types of the combined intervention, i.e., HF + PIO + F and HF+ MSDC + F, exerted a significant stimulatory effect compared with the HF mice (≈ 1.5 and ≈1.6-fold, respectively) resulting in a substantial rescue of the fractional rate of TAG synthesis (Figure 4B).

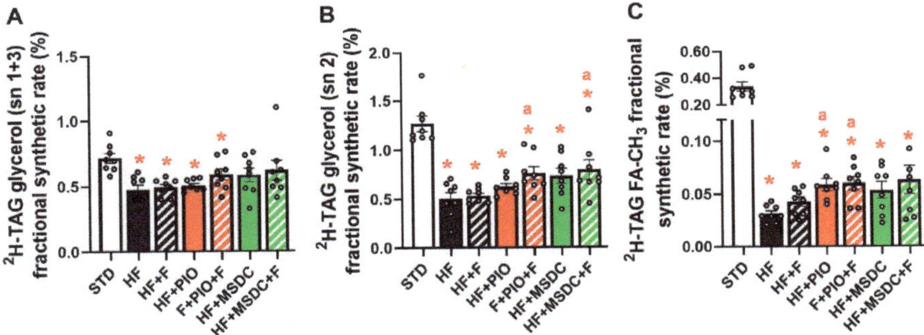

Figure 4. TAG synthesis and DNL in eWAT in vivo. Measurement of TAG synthesis (**A,B**) and DNL (FA de novo synthesis; (**C**)) in eWAT was performed using in vivo ^2H_2O incorporation in TAG in mice fed STD or various HF diets for 8 weeks. Analysis of total ^2H enrichment in sn 1 + 3 (**A**) and sn 2 (**B**) position of glycerol, and in FA methyl moieties of TAG (**C**), was performed using NMR spectroscopy. Data are means ± SEMs (*n* = 8). * Significantly different from STD (*p* ≤ 0.05, *t*-test). a—significantly different vs. HF (*p* ≤ 0.05, one-way ANOVA). TAG: triacylglycerol; DNL: de novo lipogenesis; NMR: nuclear magnetic resonance. For designation of dietary groups, see Section 2.1.

The ^2H enrichment of FA methyl moieties of TAG (DNL activity) was suppressed ≈10.8-fold by HF diet compared to STD (Figure 4B). This decrease tended to be counteracted by HF + F, HF + MSDC and HF + MSDC + F, whereas HF + PIO and HF + PIO + F exerted a significant stimulatory effect compared with the HF mice (≈1.9-fold). However, even in this case, DNL activity remained much lower as compared with the STD mice (Figure 4B).

Using LC–MS, in total 62 deuterated TAG species could be unequivocally detected in eWAT (Figure 5). Compared with the HF mice, their ^2H enrichment differed between the intervention groups. The number of differentially enriched TAG species increased in the following order: HF + PIO < HF + PIO + F < HF + F < HF + MSDC < HF + MSDC + F < STD. These data documented the most pronounced remodeling of TAG in response to the combined intervention using Omega-3 and MSDC-0602K. The above data suggest additive stimulation of TAG/FA cycling in response to both combined interventions, with a more pronounced effect of HF + MSDC + F.

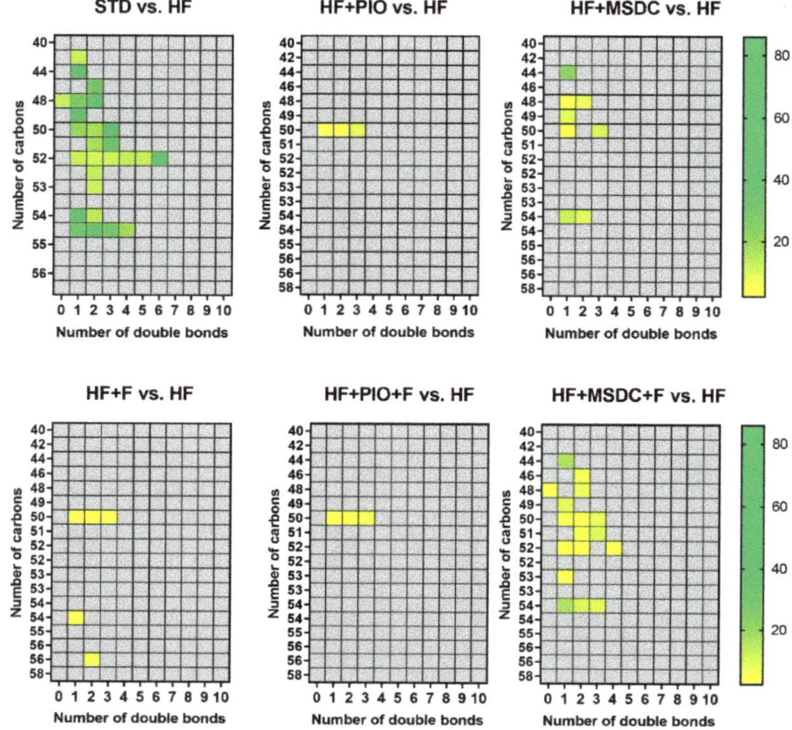

Figure 5. ^{2}H enrichment of various TAG species. In total, 62 deuterated TAG species detected using LC–MS analysis were considered. The ratio between mean enrichment of TAG in the respective intervention group and the HF mice (see the colored bar) was plotted only if the difference between groups was significant ($p \leq 0.05$, t-test). Data are plotted according to the length and saturation of TAG. Data are means $\pm$ SEMs ($n = 8$). For designation of dietary groups, see Section 2.1.

The LC–MS approach allowed for characterization of ^{2}H enrichment of the individual FA moieties in TAG (Supplementary Table S1). Compared to STD mice, the enrichment of all eight FAs considered was several times lower in mice fed HF-based diets. When comparing various interventions in mice fed HF-based diets, only HF + MSDC + F exhibited a significant effect with FA 16:0 (palmitate; ≈2.0-fold increase vs. HF, with HF + PIO + F exhibiting almost the same effect). Additionally, with FA 18:0 (oleate), HF + MSDC + F exhibited a nearly significant stimulatory effect. These data are in agreement with the relatively strong effects of both combined interventions on TAG/FA cycling in eWAT.

Eventually, peak heights of selected deuterated monosaturated to saturated FA (i.e., FA desaturation index) were evaluated as a proxy for the activity of stearoyl-CoA desaturase (**SCD**) in the tissue (Figure 6) [46,47]. In all these cases (FA 14:1/14:0, FA 16:1/16:0, FA 18:1/18:0), the index tended to be decreased by HF diet, especially when Omega-3 was admixed to HF diet, with a significant suppression by HF + F vs. HF diet in the case of 18:1/18:0 index. As a single intervention, both TZDs neutralized the effect of HF diet. However, this positive effect of TZDs was prevented in the presence of Omega-3 in the diet. These results are consistent with the stimulation of DNL by both TZDs tested, and inhibition of DNL of some FA species by Omega-3 (see Discussion).

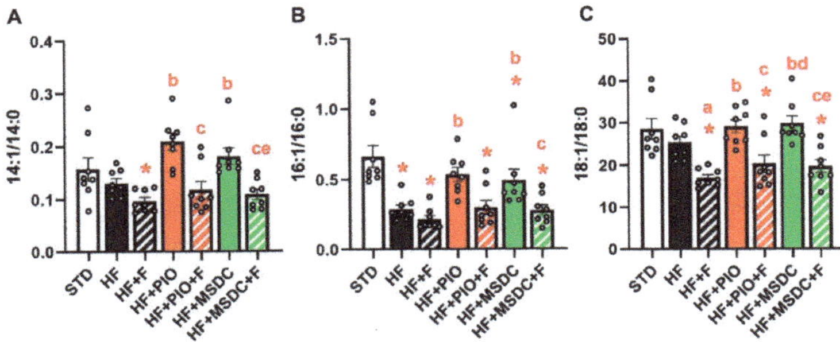

Figure 6. FA desaturation index in TAG. The ratios of the product of the desaturation and of the parent FA were calculated for myristic vs. myristoleic acid (14:1/14:0; **A**), palmitic vs. palmitoleic acid (16:1/16:0; **B**) and stearic vs. oleic acid (18:1/18:0; **C**). Data are means ± SEMs ($n = 8$). * Significantly different from STD ($p \leq 0.05$, *t*-test). a—significantly different vs. HF; b—significantly different vs. HF + F; c—significantly different vs. HF + PIO; d—significantly different vs. HF + PIO + F; e—significantly different vs. HF + MSDC ($p \leq 0.05$, one-way ANOVA). For designation of dietary groups, see Section 2.1.

3.4. Gene Expression in Adipose Tissue

Next, expression of 25 selected gene markers of regulatory pathways and metabolism in eWAT (reviewed in [2]) was evaluated using RT-qPCR. First, PLS-DA, a supervised classification method, was performed to obtain a global view on the effects of the interventions. In order to unmask possible differential effects of pioglitazone and MSDC-0602K (i) the STD mice were not considered, and (ii) the analysis was performed using two different subsets of mice. The analysis including the HF, HF + F, HF + PIO and HF + PIO + F mice (Figure 7A) revealed a separation within component 1, which described 45% of the total variation between the HF mice and two groups of mice fed PIO-containing diets (HF + PIO and HF + PIO + F), whereas Omega-3 had no effect. Similarly, the analysis focused on the potency of MSDC-0602K to modulate eWAT gene expression (Figure 7B) revealed a strong separation within component 1, which described 71% of the total variation between the HF mice and mice fed the two MSDC-containing diets (HF + MSDC and HF + MSDC + F), and a weak global effect of Omega-3 (HF vs. HF + F, and HF + MSDC vs. HF + MSDC + F, respectively). Therefore, both TZDs, pioglitazone and MSDC-0602K, exerted fundamental effects on eWAT gene expression, whereas the effect of MSDC-0602K was even more pronounced (compare Figure 7A,B). Moreover, this global analysis failed to reveal the major effect of Omega-3 on eWAT gene expression. Variable importance in projection (VIP) indicated that expression of the gene for PEPCK (*Pck1*) was the most discriminative factor within both types of analyses (compare Figure 7C,D).

Next, expression of the gene markers (reviewed in [2,3]) was analyzed in eWAT in detail (Figure 8A–H). Expression of the gene encoding PPARγ (*Pparg*; for the abbreviations and gene names, see Table 1), the key transcription factor promoting differentiation of adipocytes and regulating their lipid metabolism, was downregulated by HF diet, and it was not affected by any intervention. The expression of the mitochondrial biogenesis-inducing PGC-1α gene (*Pgc1a/Ppargc1a*), which is a target of PPARγ was also reduced by HF diet. However, it was upregulated by HF + PIO, and even more by both HF + MSDC and HF + MSDC + F (Figure 8A).

Regarding FA uptake to WAT cells, expression of the genes for LPL but not CD36 (*Lpl* and *Cd36*, respectively) was downregulated by HF diet. Expression of both genes was elevated by HF + MSDC + F, whereas it remained unaffected by the other interventions (except for *Cd36* in the HF + MSDC mice; Figure 8B).

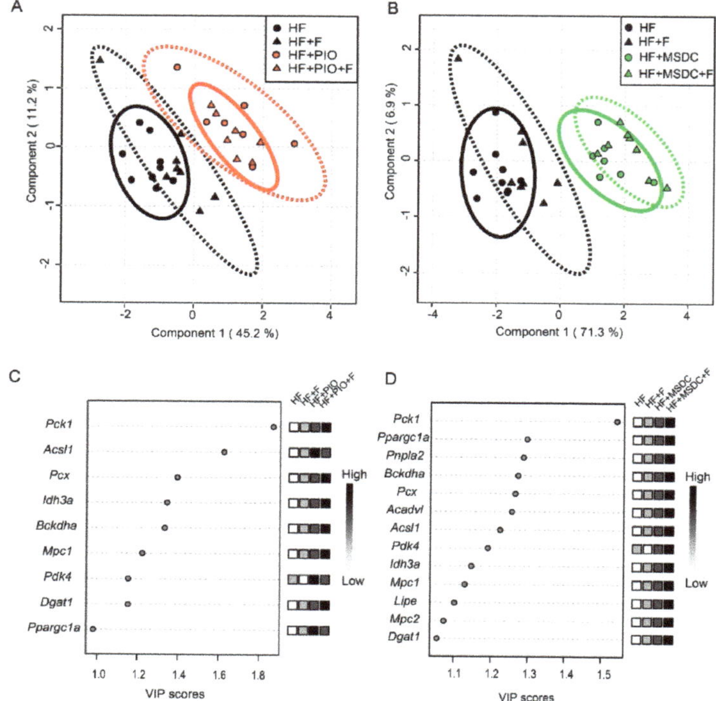

Figure 7. Multivariate analysis of gene expression in eWAT. Mice were fed various HF diets for 8 weeks. Expression of a total of 24 selected genes in eWAT was analyzed using PLS-DA (only genes with the expression level above background in all the groups were considered—i.e., all the genes in Figure 8, except for *Ucp1*. Score plots resulting from the analysis focus on the separation of the HF, HF + F, HF + PIO and HF + PIO + F mice (**A**), and the HF, HF + F, HF + MSDC and HF + MSDC + F mice (**B**). The corresponding variable importance in projection (VIP) plot with scores for identifying the most discriminating transcripts, with VIP > 1.0, is shown in (**C**) for (**A**), and in (**D**) for (**B**). The colored boxes indicate the relative expressions of the genes in each group. For designation of dietary groups, see Section 2.1.

Expression of the genes of the key lipases ATGL and HSL (*Atgl*/*Pnpla2* and *Hsl*/*Lipe*, respectively) was downregulated by HF diet, and upregulated by most of the interventions, with the additive effect of HF + MSDC + F but not HF + PIO + F resulting in the former case in a higher expression than in the STD mice (Figure 8C).

Expression of the genes involved in FA synthesis (i.e., DNL), namely, FA synthase (*Fas*/*Fasn*) and FA elongase 5 (*Elovl5*), was strongly downregulated by HF diet, and it was not affected by any intervention (vs. HF diet; Figure 8D).

Among the markers of glyceroneogenesis and FA re-esterification (Figure 8E), expression of the gene for GK (*Gk*) was upregulated by HF diet, and it was also increased by both TZDs, but not by Omega-3. Expression of the genes engaged in glyceroneogenesis in WAT, namely, PC (*Pc*/*Pcx*) and PEPCK (*Pck1*), was strongly downregulated by HF and upregulated by both TZDs. The stimulatory effect of MSDC-0602K was more pronounced, with the combined HF + MSDC + F (but not HF + PIO + F) intervention showing an additive effect. Expression of the genes for DGAT1 and DGAT2 (*Dgat1* and *Dgat2*, respectively), involved in FA re-esterification [48], was suppressed by HF diet. Expression of *Dgat1* was upregulated by all the interventions, except for HF + F, with the combined HF + MSDC + F (but not HF + PIO + F) intervention showing an additive effect. Expression of *Dgat2* was only affected by the HF + MSDC + F, which normalized the expression to the level in STD group.

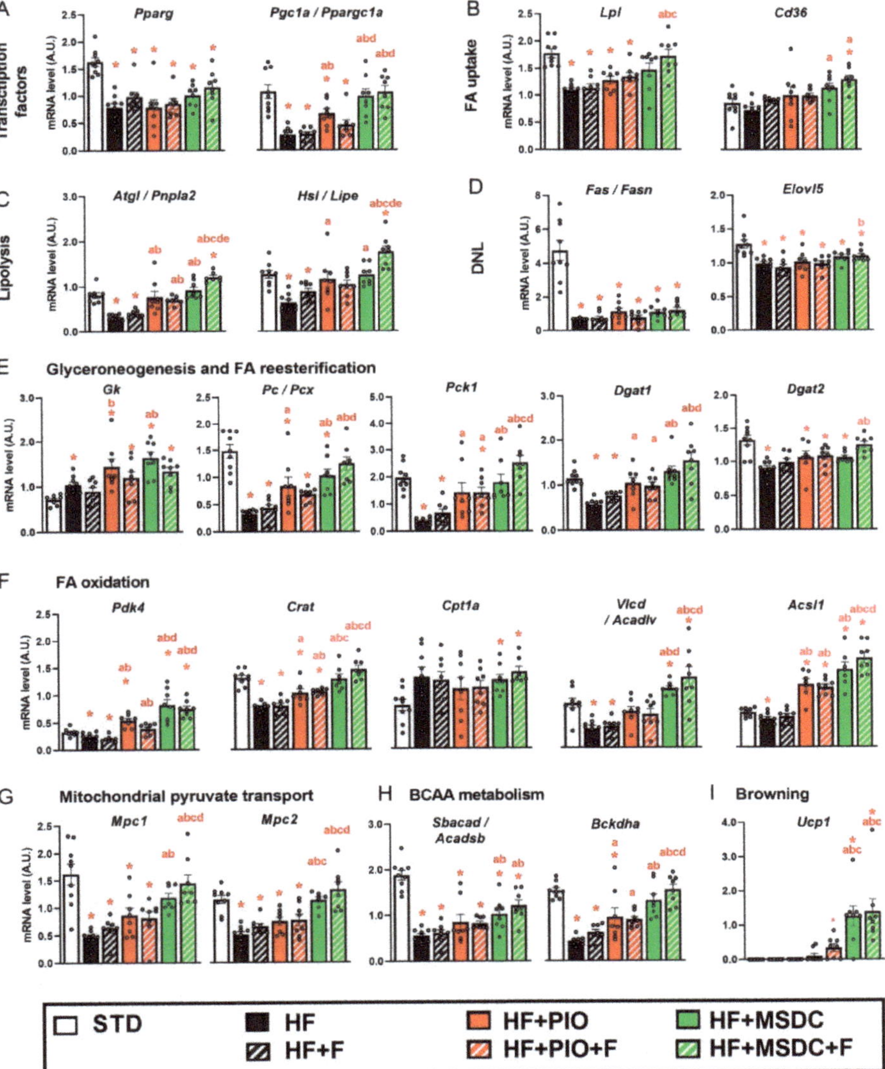

Figure 8. Expression of selected genes in eWAT. Mice were fed STD or various HF diets for 8 weeks. Expressions of the genes engaged in various regulatory and metabolic pathways (**A–I**) were evaluated using RT-qPCR and normalized to the mean signal of four reference genes (see Materials and Methods). Data are means ± SEMs ($n = 8$). * Significantly different from STD ($p \leq 0.05$, *t*-test). a—significantly different vs. HF; b—significantly different vs. HF + F; c—significantly different vs. HF + PIO; d—significantly different vs. HF + PIO + F; e—significantly different vs. HF + MSDC ($p \leq 0.05$, one-way ANOVA. For designation of dietary groups, see Section 2.1.

Regarding FA oxidation (Figure 8F), expression of the gene for PDK4 (*Pdk4*), which limits glucose oxidation by inhibiting pyruvate dehydrogenase and thus supports β-oxidation [49], was downregulated by HF diet. It was increased by both TZDs, with a stronger effect of MSDC-0602K, and it was insensitive to Omega-3. A very similar pattern was observed with the gene for CRAT (*Crat*), which acts in concert with PDK4 to regulate mitochondrial fuel oxidation, and which is associated with insulin sensitivity [50]. Expression of CPT1α gene (*Cpt1a*), encoding protein essential for the transport of FAs for their β-oxidation in mitochondria, was upregulated by HF diet, and the expression was similar across all the interventions. Expression of gene for VLCAD (*Vlcad/Acadvl*), enzyme catalyzing the first step of mitochondrial β-oxidation was downregulated by the HF diet, and it was upregulated by the two interventions containing MSDC-0602K. *Vlcad/Acadvl* gene expression in both HF + MSDC and HF + MSDC + F was even higher compared to the STD mice. However, *Vlcad/Acadvl* expression was unaffected by pioglitazone. The expression of gene for ACSL1 (*Acsl1*), which converts free long-chain FA into fatty acyl-CoA esters, was only marginally downregulated by the HF diet; it was not affected by HF + F. It was strongly upregulated by both TZDs, with the most pronounced effect being in HF + MSDC + F mice.

Expression of the genes for MPC1 and MPC2 (Figure 8G), the two forms of mitochondrial pyruvate carrier (*Mpc1* and *Mpc2*; [51]), was downregulated by HF diet. It was induced by MSDC-0602K (but not pioglitazone), independently of Omega-3.

The gene screen data above document a general decrease in the metabolic activities of eWAT in response to HF diet. This was apparent across the gene markers of all the metabolic pathways studied (except for *Gk*, *Cd36* and *Cpt1a*). Although both tested TZDs ameliorated the inhibitory effect of HF diet on the expression of most genes, MSDC-0602K exerted a stronger effect. Both TZDs promoted lipolysis and FA re-esterification, in association with the elevation of mitochondrial β-oxidation. The induction of FA oxidation was documented using the measurements of ^{14}C-palmitate oxidation in fragments of eWAT dissected from the mice at the end of the dietary interventions (Supplementary Figure S1). It was in concert with the induction of mitochondrial biogenesis, documented by the increase in both the *Pgc1a/Ppargc1a* expression and the specific content of the complex III of mitochondrial respiratory chain in eWAT (Supplementary Figure S2). FA re-esterification was supported by enhanced activity of glyceroneogenesis mediated by PC and PEPCK, one of the most affected pathways by the studied interventions, but not by DNL. This indicated that induction of glycerol 3-phosphate by various interventions (see above) resulted mainly from glyceroneogenesis (see Discussion).

In general, Omega-3 tended to augment the stimulatory effect of MSDC-0602K, and this combination even exerted a significant additive effect on lipolysis-related genes (*Atgl/Pnpla2* and *Hsl/Lipe*). In contrast, Omega-3 tended to counteract the stimulatory effects of pioglitazone, with a significant influence at the level of PDK4. Pyruvate transport to mitochondria could be induced by MSDC-0602K, but not by pioglitazone or Omega-3.

Next, we focused on the expression of the genes for enzymes engaged in metabolism of branched-chain amino acids (BCAA), due to the role of BCAA metabolism in WAT in whole-body insulin sensitivity [52,53] and its improvement by TZDs [54]. Expression of these genes, namely, *Sbacad/Acadsb* and *Bckdha*, was downregulated by HF diet (Figure 8H). While pioglitazone increased the expression of *Bckdha*, MSDC-0602K stimulated the expression of both genes more than pioglitazone. Even in this case, Omega-3 supplementation (i.e., the HF + F diet) had no significant effect, but it tended to augment the stimulatory effect of MSDC-0602K. These data are consistent with the involvement of BCAA metabolism in WAT of diet-induced obese mice in the prevention of insulin resistance by both TZDs.

Lastly, expression of the gene for UCP1, which mediates thermogenesis in both brown and brite adipocytes (*Ucp1*; reviewed in [1]), was evaluated (Figure 8I). No expression above the background could be observed in the STD, HF or HF + F mice. In the remaining groups, *Ucp1* was expression was induced to various extents, with both MSDC-0602K-containing diets exerting an equal effect, higher compared with the other diets.

In order to learn whether the above described effects of various dietary interventions on gene expression in eWAT depended on the anatomical location of the tissue, expression levels of several of the studied genes were also evaluated in scWAT of the same animals (Figure 9). The gene expression pattern was very similar in both WAT depots compared. Only *Ucp1* expression showed fat depot-specific differences, namely, a remarkably scattered expression in scWAT of the STD-fed mice (Figure 9), which contrasted with the lack of *Ucp1* expression in eWAT of these animals (Figure 8I). Feeding HF diet eliminated *Ucp1* expression in scWAT, while the two MSDC-0602K-containing diets rescued the expression to the levels observed in the STD mice (Figure 8I). When the relative levels of *Ucp1* expression were compared across eWAT, scWAT and interscapular brown fat, the highest levels found in the two WAT depots were not significantly different, and they were two orders of magnitude lower compared with those in brown adipose tissue of these mice (not shown).

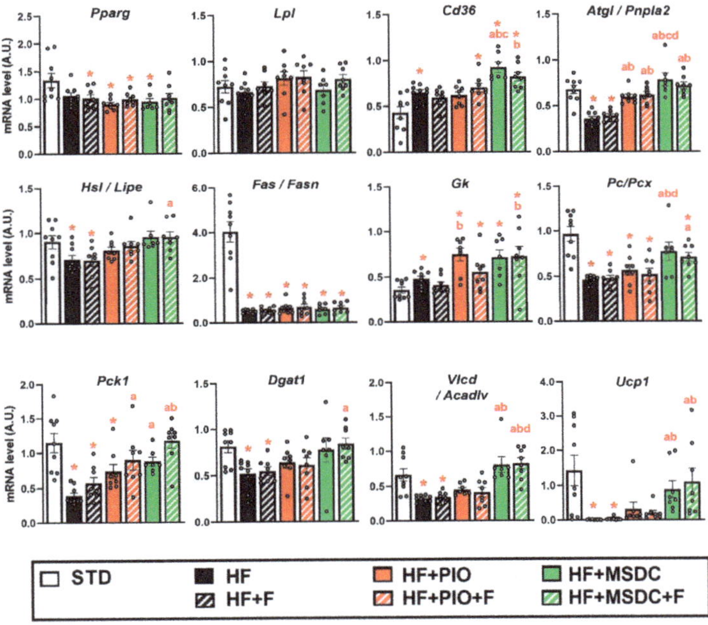

Figure 9. Expression of selected genes in subcutaneous WAT. Mice were fed STD or various HF diets for 8 weeks. Expression of the subset of genes shown in Figure 8 was assessed using RT-qPCR and normalized to the mean signal of four reference genes (see Materials and Methods). Data are means ± SEMs (*n* = 8). * significantly different from STD (*p* ≤ 0.05, *t*-test). a—significantly different vs. HF; b—significantly different vs. HF + F; c—significantly different vs. HF + PIO; d—significantly different vs. HF + PIO + F. For designation of dietary groups, see Section 2.1.

4. Discussion

The principal finding of this report was that additive beneficial effects of Omega-3 and TZDs in mice fed obesogenic HF diet were associated with pronounced changes in gene expression (Figure 8) and metabolism of eWAT. These effects included the reduction in body weight gain and adiposity (Figure 1); preservation of glucose homeostasis and dyslipidemia (Figure 3 and Table 2); and reduction in liver fat content (Table 2) and low-grade inflammation of WAT (Figure 2C). In terms of WAT function, the results suggest the role of TAG/FA cycling (Figure 10) in the beneficial effects of the studied interventions.

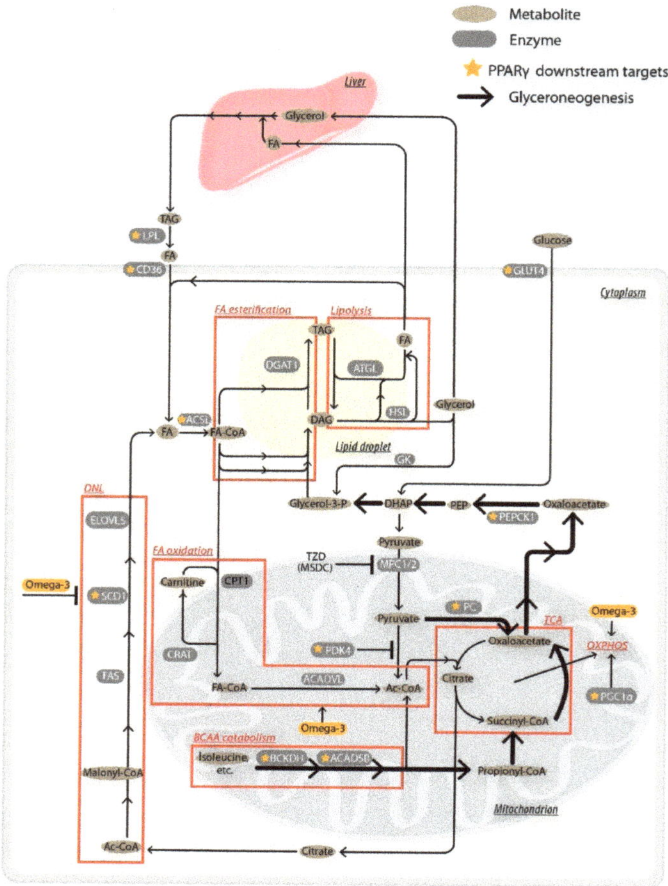

Figure 10. Hypothetical model for the complex mechanism of the induction of TAG/FA cycling activity in WAT by the combined intervention using Omega-3 and TZDs in adipocytes. Both pioglitazone and MSDC-0602K stimulate transcriptional factor PPARγ and its targets, resulting in increased glucose and FA uptake into the adipocyte. In parallel, mitochondrial FA oxidation is upregulated compared with glucose oxidation, reflecting (i) higher FA influx; (ii) higher Omega-3 mediated induction of genes involved in FA oxidation, mitochondrial biogenesis and OXPHOS activity; (iii) inhibition of pyruvate transport to mitochondria—especially in response to MSDC-0602K; and (iv) induction of PDK4, which results in inhibition of oxidation of a limited amount of pyruvate entering mitochondria. Thus, synthesis of glycerol-3-phosphate (glycerol-3-P) is increased by both conversion from glucose and glyceroneogenesis from pyruvate. Induction of BCAA catabolism provides another source of glyceroneogenic substrate oxaloacetate. The increased levels of glycerol-3-P and FA-CoA stimulate synthesis of TAG, linked to concomitant increase in lipolysis. FA released from TAG can be either oxidized or re-esterified (in TAG/FA cycle). The re-esterification can occur in adipocytes (primary FA re-esterification) or extra-adipose tissues (secondary FA re-esterification), mainly the liver (reviewed in [29,32]). Glycerol released during lipolysis is transported via blood to the liver to support (i) gluconeogenesis, and (ii) the formation of glycerol-3-P for TAG synthesis, either from glucose, or via direct phosphorylation of glycerol. For other details, references and abbreviations, see the main text and Table 1. For the explanation of the pleiotropic effects of TZDs, see also [22]. Ac—acetyl; CoA—coenzyme A; G3P—glyceraldehyde-3-phosphate; LPA—lysophosphatidic acid; P—phosphate.

To our knowledge, this is the first study to characterize the direct effects of the combined intervention using Omega-3 and TZDs on WAT metabolism. Our results document a decrease of metabolic activities in eWAT by HF diet [55], and a partial counteraction of this effect by various interventions. Thus, the measurements of glyceroneogenesis using in vivo labeling of TAG by deuterium from ^2H_2O, and subsequent analyses of TAG composition using NMR spectroscopy and LC–MS, documented an additive interaction between Omega-3 and both types of TZDs in the stimulation of this activity. FA re-esterification was stimulated the most by HF + MSDC + F, as documented by the highest number of deuterated TAG species (detected using LC–MS) in response to this intervention vs. HF diet. Additive stimulation by Omega-3 and TZDs, with a stronger effect of HF + MSDC + F, was observed in both eWAT and scWAT, as documented by the expression of the genes involved in FA uptake (*Lpl* and *Cd36*), lipolysis (*Atgl/Pnpla2* and *Hsl/Lipe*), glyceroneogenesis (*Pc/Pcx* and *Pck1*), FA re-esterification (*Dgat1*) and FA oxidation (*Crat*, *Vlcd/Acadlv* and *Acsl1*), and the biochemical data. The changes in gene expression suggested that stimulation of glyceroneogenesis depended more on PC-PEPCK-mediated glycerol 3-phosphate formation rather than GK, which is also consistent with the preferential activation of the earlier pathway [29,38] and the induction of *Pdk4* [35] by TZD. However, the relatively low upregulation of *Gk* expression, which was also observed in both eWAT and scWAT in response to various interventions, suggested that direct phosphorylation of glycerol was also involved in the increased formation of glycerol 3-phosphate. Thus, the activation of TAG/FA cycling in response to various interventions could be even higher than illustrated by the measurements of glyceroneogenesis using labeling of TAG by deuterium from ^2H_2O (see above). That *Dgat1* but not *Dgat2* expression was affected reflects (i) the interaction of DGAT1 with ATGL during lipolysis and FA re-esterification, and (ii) the involvement of DGAT2 in the esterification of FA formed by DNL [48]. Collectively, the data document the additive stimulation of TAG/FA cycling by Omega-3 and TZDs, especially by the combined intervention that involved MSDC-0602K.

Direct measurement of DNL in vivo in eWAT and the corresponding gene expression analysis (*Fas/Fasn*) indicated very strong suppression of DNL by HF diet, and a limited rescue of this activity by both TZDs [56]. Moreover, the effect of HF diet was further augmented by Omega-3 [57]; these FAs even prevented activation of DNL by TZDs. These results were further supported by the changes in the FA desaturation index, marking SCD activity and lipogenesis in the tissue [46,47]. Interestingly, stimulation of TAG/FA cycling in WAT by prolonged cold exposure of mice was accompanied by a pronounced stimulation of DNL in the tissue [33], documenting independent regulation of glyceroneogenesis (TAG/FA cycling) and DNL in WAT.

Molecular mechanisms of the combined intervention on eWAT gene expression and metabolism probably involve interactions of the tested compounds with intracellular regulatory pathways in adipocytes, including PPARs, endocannabinoids, AMP-activated protein kinase and others (reviewed in [1,2,22,33]). These Omega-3-induced changes result, namely, in increased capacity for FA oxidation in the mitochondria, linked with OXPHOS (Figure 10). Some effects in WAT may be secondary to the activation of lipid metabolism in the liver and other tissues, especially due to the PPARα-mediated response to Omega-3 [58]. The effects of TZDs are even more pleiotropic, reflecting the differential activation of multiple PPARγ targets, and the reversible inhibition of pyruvate transport into mitochondria ([25–27]; Figure 10). Despite the additive stimulation by Omega-3 and TZDs of most of the metabolic activities of WAT, which are the basis of the "healthy adipocyte" phenotype (reviewed in [1–3]), some metabolic pathways were affected differently. Thus, DNL is inhibited by Omega-3 [57] and stimulated by TZDs ([56]; see above), reflecting probably the differential control via SCD1 [47,59].

Moreover, stimulation of glucose uptake due to the insulin-sensitizing effect of TZDs probably plays a major role in the stimulation of DNL. Expression of the genes of BCAA metabolism was upregulated by TZDs [54], independent of Omega-3, in agreement with the role of BCAA in the insulin-sensitizing effect of TZDs [52,53]. Overall, the data suggest a complex modulation of WAT metabolic activities, which converge in the TAG/FA cycle. The most pronounced effects were produced by the combination of HF + MSDC + F. This may indicate that the modulation of pyruvate's entry into mitochondria by MSDC-0602K has a major impact on these processes [25–27]. As shown graphically in Figure 10, the effects observed in WAT in this study involve mitochondrial and extramitochondrial metabolism in both the adipose tissue and the liver. The Omega-3 and TZD interactions can occur at various regulatory levels and may also involve both direct and downstream modulation of PPARγ, and likely other transcriptional networks (Figure 10).

Our results support the role of TAG/FA cycling in WAT in a healthy metabolic phenotype. This core biochemical activity of WAT is essential for flexible control of plasma NEFA levels and supply of FA to serve as energy fuel [1,29–32]. Insufficient capacity of TAG/FA cycling in WAT probably underlies insufficient "expandability" of WAT [60], i.e., pathological low capacity of this tissue to serve as a buffer for circulating FAs. This leads to lipotoxic damage to metabolism and insulin sensitivity in extra-adipose tissues. Indeed, in WAT of obese [61] and diabetic patients [62], the activities of both TAG/FA cycling and DNL are relatively low. Additionally, OXPHOS slows down in obesity and with aging [63], whereas it is increased by exercise [64]. Thus, the additive stimulatory effect of TZDs and Omega-3 on TAG/FA cycling in WAT could provide an important determinant of metabolic health.

The experimental design was based on the doses of Omega-3 and TZDs used in the previous studies on HF diet-fed mice (see Section 2.1). Regarding the TZDs, when re-calculated to human equivalent dose (HED; [65]), 50 mg pioglitazone/kg of the HF + PIO or HF + PIO + F diet was equivalent to 32 mg HED (with 15–45 mg pioglitazone/day, the usual therapeutic dose in diabetic patients), whereas 331 mg MSDC-0602K/kg of the HF + MSDC or HF + MSDC + F diet was equivalent to 146 mg HED (with 60–250 mg MSDC-0602K/day used in a clinical trial [28]). The ratio between the dose of MSDC-0602K and pioglitazone, i.e., ≈6.6, was relatively low with respect to the ≈16-fold lower affinity of MSDC-0602K to PPARγ as compared with pioglitazone [25]. Nevertheless, the effects of MSDC-0602K were mostly stronger compared with pioglitazone. These differences were especially pronounced in the combined interventions HF + MSDC + F and HF + PIO + F (see below). The relatively high potency of MSDC-0602K was following the notion that, unlike pioglitazone, the effects of MSDC-0602K depended largely on the direct inhibition of mitochondrial pyruvate carriers MPC1 and MPC2, rather than on the activation of PPARγ-mediated transcription [25–27]. Indeed, expression of *Mpc1* and *Mpc2*, which was downregulated by HF diet, was increased by HF + MSDC and HF + MSDC + F, but not by any other intervention. This suggests an adaptive response at the level of *Mpc1* and *Mpc2* transcription.

The effect of the interventions on gross energy balance (i.e., lower adiposity accretion in face of equal energy intake in all the HF-based diets compared) needs to be clarified, especially for diets containing MSDC-0602K. The present data could not rule out a possibility that a decrease in dietary energy intake in the gut was involved. Consistent with the previous studies (reviewed in [1]), we have observed a small increase in *Ucp1* expression in WAT, reaching much lower levels than in BAT. Thus, UCP1-mediated thermogenesis could be involved, but its characterization was out of the scope of this study.

Regarding the involvement of energy expenditure in the anti-obesity effects of the studied interventions, a seminal recent study should be recalled [31]. It highlighted the role of futile metabolic cycling of circulating metabolites for a steady supply of appropriate nutrients to each tissue, despite varying dietary intake. It demonstrated that (i) a majority of circulating carbon flux reflects glucose-lactate and TAG-glycerol-FA cycling, and (ii) lactate and FA are the two major energy fuels. The rate of this futile cycling was much faster [31] than so far assumed [29,32]. As shown before [32], the flux of circulating FA depends almost entirely on TAG/FA cycling in WAT. Thus, the energy requirements for TAG/FA cycling in WAT, associated with futile cycling of circulating FA and glycerol, must be relatively high. This energy expenditure should affect significantly whole-body energy balance. This could largely explain the anti-obesity effects of various interventions. Indeed, the HF + MSDC + F intervention that resulted in the highest stimulation of TAG turnover in eWAT exerted the most pronounced reduction of HF diet-induced obesity. Accordingly, as we have shown recently in the obesity-resistant A/J and obesity-prone B6 mice exposed to cold stress [33], obesity resistance is associated with higher stimulability of TAG/FA cycling in eWAT.

5. Conclusions

Modulation of intrinsic metabolic features of WAT, and in particular stimulating the level of TAG/FA cycling, is involved in additive beneficial effects of Omega-3 and TZDs on metabolic health in diet-induced obese mice. Whereas TAG/FA cycling in WAT is suppressed by obesogenic HF diet, its partial rescue linked to the increased flux of circulating metabolites could reduce obesity. WAT metabolism represents an important target for treatment strategies for obesity and associated diseases. Combined interventions using Omega-3 could both augment the positive effects of TZD insulin sensitizers and minimize/prevent the weight gain that can occur with this pharmacotherapy. The second-generation PPARγ-sparing TZDs represent prospective potent pharmaceuticals for use in the combined intervention.

Supplementary Materials: The following are available online at http://www.mdpi.com/2072-6643/12/12/3737/s1. Method:LC–MS-based lipidomics. Figure S1: Palmitate oxidation in eWAT. Figure S2: Quantification of the complexes of mitochondrial respiratory chain in eWAT. Table S1: ^{2}H enrichment of various FA moieties in response to the interventions.

Author Contributions: Conceptualization, P.F., M.R., J.C. and J.K.; writing—original draft, K.B. and J.K.; writing—review and editing, K.B., J.F., O.H., P.Z., M.R., J.C. and J.K.; investigation, K.B., J.F., R.P., T.C., O.H., I.I., M.H., P.J. and P.F.; methodology, K.B., R.P., T.C., O.K., M.H. and P.F.; formal analysis, K.B., J.F., T.C., O.K., O.H., K.A., M.H., P.F., L.L., P.Z. and M.R.; visualization, K.B., J.F., O.H., K.A. and P.Z.; project administration, K.B., P.F. and J.K.; resources, J.C.; validation, K.B.; data curation, K.B. and T.C.; funding acquisition, J.K.; supervision, J.K. All authors have read and agreed to the published version of the manuscript.

Funding: This research was funded by the Czech Science Foundation (grant No. 19-02411S).

Acknowledgments: We thank Sona Hornova, Karolina Sedova and Daniela Salkova for their excellent technical assistance and Sara Stanić for her help with the illustrations. Omega-3 PUFA concentrate and MSDC-0602K were provided by Epax (Ålesund, Norway) and Cirius Therapeutics (Kalamazoo, USA), respectively.

Conflicts of Interest: J.C. is a cofounder, employee and owns stock in Cirius Therapeutics, which owns and is undertaking clinical development of MSDC-0602K.

References

1. Flachs, P.; Rossmeisl, M.; Kuda, O.; Kopecky, J. Stimulation of mitochondrial oxidative capacity in white fat independent of UCP1: A key to lean phenotype. *Biochim. Biophys. Acta (BBA) Mol. Cell Biol. Lipids* **2013**, *1831*, 986–1003. [CrossRef] [PubMed]

2. Masoodi, M.; Kuda, O.; Rossmeisl, M.; Flachs, P.; Kopecky, J. Lipid signaling in adipose tissue: Connecting inflammation & metabolism. *Biochim. Biophys. Acta (BBA) Mol. Cell Biol. Lipids* **2015**, *1851*, 503–518. [CrossRef]

3. Kuda, O.; Rossmeisl, M.; Kopecky, J. Omega-3 fatty acids and adipose tissue biology. *Mol. Asp. Med.* **2018**, *64*, 147–160. [CrossRef] [PubMed]

4. Lai, H.T.; de Oliveira Otto, M.C.; Lemaitre, R.N.; McKnight, B.; Song, X.; King, I.B.; Chaves, P.H.; Odden, M.C.; Newman, A.B.; Siscovick, D.S.; et al. Serial circulating omega 3 polyunsaturated fatty acids and healthy ageing among older adults in the Cardiovascular Health Study: Prospective cohort study. *BMJ* **2018**, *363*, k4067. [CrossRef]

5. Manson, J.E.; Cook, N.R.; Lee, I.M.; Christen, W.; Bassuk, S.S.; Mora, S.; Gibson, H.; Albert, C.M.; Gordon, D.; Copeland, T.; et al. Marine n−3 Fatty Acids and Prevention of Cardiovascular Disease and Cancer. *N. Engl. J. Med.* **2018**, *380*, 23–32. [CrossRef]

6. Flachs, P.; Ruhl, R.; Hensler, M.; Janovska, P.; Zouhar, P.; Kus, V.; Macek, J.Z.; Papp, E.; Kuda, O.; Svobodova, M.; et al. Synergistic induction of lipid catabolism and anti-inflammatory lipids in white fat of dietary obese mice in response to calorie restriction and n-3 fatty acids. *Diabetologia* **2011**, *54*, 2626–2638. [CrossRef] [PubMed]

7. Rossmeisl, M.; Medrikova, D.; van Schothorst, E.M.; Pavlisova, J.; Kuda, O.; Hensler, M.; Bardova, K.; Flachs, P.; Stankova, B.; Vecka, M.; et al. Omega-3 phospholipids from fish suppress hepatic steatosis by integrated inhibition of biosynthetic pathways in dietary obese mice. *Biochim. Biophys. Acta Mol. Cell Biol. Lipids* **2014**, *1841*, 267–278. [CrossRef] [PubMed]

8. de Castro, G.S.; Calder, P.C. Non-alcoholic fatty liver disease and its treatment with n-3 polyunsaturated fatty acids. *Clin. Nutr.* **2018**, *37*, 37–55. [CrossRef]

9. Flachs, P.; Mohamed-Ali, V.; Horakova, O.; Rossmeisl, M.; Hosseinzadeh-Attar, M.J.; Hensler, M.; Ruzickova, J.; Kopecky, J. Polyunsaturated fatty acids of marine origin induce adiponectin in mice fed high-fat diet. *Diabetologia* **2006**, *49*, 394–397. [CrossRef] [PubMed]

10. Wu, J.H.; Cahill, L.E.; Mozaffarian, D. Effect of fish oil on circulating adiponectin: A systematic review and meta-analysis of randomized controlled trials. *J. Clin. Endocrinol. Metab.* **2013**, *98*, 2451–2459. [CrossRef]

11. Calder, P.C. Marine omega-3 fatty acids and inflammatory processes: Effects, mechanisms and clinical relevance. *Biochim. Biophys. Acta (BBA) Mol. Cell Biol. Lipids* **2015**, *1851*, 469–484. [CrossRef] [PubMed]

12. van Schothorst, E.M.; Flachs, P.; Franssen-van Hal, N.L.; Kuda, O.; Bunschoten, A.; Molthoff, J.; Vink, C.; Hooiveld, G.J.; Kopecky, J.; Keijer, J. Induction of lipid oxidation by polyunsaturated fatty acids of marine origin in small intestine of mice fed a high-fat diet. *BMC Genom.* **2009**, *10*, 110. [CrossRef] [PubMed]

13. Kroupova, P.; van Schothorst, E.M.; Keijer, J.; Bunschoten, A.; Vodicka, M.; Irodenko, I.; Oseeva, M.; Zacek, P.; Kopecky, J.; Rossmeisl, M.; et al. Omega-3 Phospholipids from Krill Oil Enhance Intestinal Fatty Acid Oxidation More Effectively than Omega-3 Triacylglycerols in High-Fat Diet-Fed Obese Mice. *Nutrients* **2020**, *12*, 37. [CrossRef] [PubMed]

14. Ruzickova, J.; Rossmeisl, M.; Prazak, T.; Flachs, P.; Sponarova, J.; Vecka, M.; Tvrzicka, E.; Bryhn, M.; Kopecky, J. Omega-3 PUFA of marine origin limit diet-induced obesity in mice by reducing cellularity of adipose tissue. *Lipids* **2004**, *39*, 1177–1185. [CrossRef]

15. Adamcova, K.; Horakova, O.; Bardova, K.; Janovska, P.; Brezinova, M.; Kuda, O.; Rossmeisl, M.; Kopecky, J. Reduced Number of Adipose Lineage and Endothelial Cells in Epididymal fat in Response to Omega-3 PUFA in Mice Fed High-Fat Diet. *Mar. Drugs* **2018**, *16*, 515. [CrossRef]

16. Kunesova, M.; Braunerova, R.; Hlavaty, P.; Tvrzicka, E.; Stankova, B.; Skrha, J.; Hilgertova, J.; Hill, M.; Kopecky, J.; Wagenknecht, M.; et al. The influence of n-3 polyunsaturated fatty acids and very low calorie diet during a short-term weight reducing regimen on weight loss and serum fatty acid composition in severely obese women. *Physiol. Res.* **2006**, *55*, 63–72.

17. Mori, T.A.; Bao, D.Q.; Burke, V.; Puddey, I.B.; Watts, G.F.; Beilin, L.J. Dietary fish as a major component of a weight-loss diet: Effect on serum lipids, glucose, and insulin metabolism in overweight hypertensive subjects. *Am. J. Clin. Nutr.* **1999**, *70*, 817–825. [CrossRef]

18. Flachs, P.; Rossmeisl, M.; Kopecky, J. The Effect of n-3 Fatty Acids on Glucose Homeostasis and Insulin Sensivity. *Physiol. Res.* **2014**, 93–118. [CrossRef]

19. Kuda, O.; Jelenik, T.; Jilkova, Z.; Flachs, P.; Rossmeisl, M.; Hensler, M.; Kazdova, L.; Ogston, N.; Baranowski, M.; Gorski, J.; et al. n-3 fatty acids and rosiglitazone improve insulin sensitivity through additive stimulatory effects on muscle glycogen synthesis in mice fed a high-fat diet. *Diabetologia* **2009**, *52*, 941–951. [CrossRef]

20. Kus, V.; Flachs, P.; Kuda, O.; Bardova, K.; Janovska, P.; Svobodova, M.; Jilkova, Z.M.; Rossmeisl, M.; Wang-Sattler, R.; Yu, Z.; et al. Unmasking Differential Effects of Rosiglitazone and Pioglitazone in the Combination Treatment with n-3 Fatty Acids in Mice Fed a High-Fat Diet. *PLoS ONE* **2011**, *6*, e27126–e27127. [CrossRef]

21. Choi, J.H.; Banks, A.S.; Estall, J.L.; Kajimura, S.; Bostrom, P.; Laznik, D.; Ruas, J.L.; Chalmers, M.J.; Kamenecka, T.M.; Bluher, M.; et al. Anti-diabetic drugs inhibit obesity-linked phosphorylation of PPARgamma by Cdk5. *Nature* **2010**, *466*, 451–456. [CrossRef] [PubMed]

22. Divakaruni, A.S.; Wiley, S.E.; Rogers, G.W.; Andreyev, A.Y.; Petrosyan, S.; Loviscach, M.; Wall, E.A.; Yadava, N.; Heuck, A.P.; Ferrick, D.A.; et al. Thiazolidinediones are acute, specific inhibitors of the mitochondrial pyruvate carrier. *Proc. Natl. Acad. Sci. USA* **2013**, *110*, 5422–5427. [CrossRef] [PubMed]

23. Colca, J.R.; McDonald, W.G.; Cavey, G.S.; Cole, S.L.; Holewa, D.D.; Brightwell-Conrad, A.S.; Wolfe, C.L.; Wheeler, J.S.; Coulter, K.R.; Kilkuskie, P.M.; et al. Identification of a mitochondrial target of thiazolidinedione insulin sensitizers (mTOT)–relationship to newly identified mitochondrial pyruvate carrier proteins. *PLoS ONE* **2013**, *8*, e61551. [CrossRef] [PubMed]

24. Wang, S.; Dougherty, E.J.; Danner, R.L. PPARgamma signaling and emerging opportunities for improved therapeutics. *Pharmacol. Res.* **2016**, *111*, 76–85. [CrossRef] [PubMed]

25. Chen, Z.; Vigueira, P.A.; Chambers, K.T.; Hall, A.M.; Mitra, M.S.; Qi, N.; McDonald, W.G.; Colca, J.R.; Kletzien, R.F.; Finck, B.N. Insulin resistance and metabolic derangements in obese mice are ameliorated by a novel peroxisome proliferator-activated receptor gamma-sparing thiazolidinedione. *J. Biol. Chem.* **2012**, *287*, 23537–23548. [CrossRef] [PubMed]

26. McCommis, K.S.; Hodges, W.T.; Brunt, E.M.; Nalbantoglu, I.; McDonald, W.G.; Holley, C.; Fujiwara, H.; Schaffer, J.E.; Colca, J.R.; Finck, B.N. Targeting the mitochondrial pyruvate carrier attenuates fibrosis in a mouse model of nonalcoholic steatohepatitis. *Hepatology* **2017**, *65*, 1543–1556. [CrossRef]

27. Colca, J.R.; Tanis, S.P.; McDonald, W.G.; Kletzien, R.F. Insulin sensitizers in 2013: New insights for the development of novel therapeutic agents to treat metabolic diseases. *Expert. Opin. Investig. Drugs* **2014**, *23*, 1–7. [CrossRef]

28. Harrison, S.A.; Alkhouri, N.; Davison, B.A.; Sanyal, A.; Edwards, C.; Colca, J.R.; Lee, B.H.; Loomba, R.; Cusi, K.; Kolterman, O.; et al. Insulin sensitizer MSDC-0602K in non-alcoholic steatohepatitis: A randomized, double-blind, placebo-controlled phase IIb study. *J. Hepatol.* **2020**, *72*, 613–626. [CrossRef]

29. Reshef, L.; Olswang, Y.; Cassuto, H.; Blum, B.; Croniger, C.M.; Kalhan, S.C.; Tilghman, S.M.; Hanson, R.W. Glyceroneogenesis and the triglyceride/fatty acid cycle. *J. Biol. Chem.* **2003**, *278*, 30413–30416. [CrossRef]

30. Newsholme, E.A.; Crabtree, B. Substrate cycles: Their metabolic energy and thermic consequences in man. *Biochem. Soc. Symp.* **1976**, *43*, 183–205.

31. Hui, S.; Cowan, A.J.; Zeng, X.; Yang, L.; TeSlaa, T.; Li, X.; Bartman, C.; Zhang, Z.; Jang, C.; Wang, L.; et al. Quantitative Fluxomics of Circulating Metabolites. *Cell. Metab.* **2020**, *32*, 676–688. [CrossRef] [PubMed]

32. Kalderon, B.; Mayorek, N.; Berry, E.; Zevit, N.; Bar-Tana, J. Fatty acid cycling in the fasting rat. *Am. J. Physiol. Endocrinol. Metab.* **2000**, *279*, E221–E227. [CrossRef] [PubMed]

33. Flachs, P.; Adamcova, K.; Zouhar, P.; Marques, C.; Janovska, P.; Viegas, I.; Jones, J.G.; Bardova, K.; Svobodova, M.; Hansikova, J.; et al. Induction of lipogenesis in white fat during cold exposure in mice: Link to lean phenotype. *Int. J. Obes.* **2017**, *41*, 372–380. [CrossRef] [PubMed]

34. Bederman, I.R.; Foy, S.; Chandramouli, V.; Alexander, J.C.; Previs, S.F. Triglyceride synthesis in epididymal adipose tissue: Contribution of glucose and non-glucose carbon sources. *J. Biol. Chem.* **2009**, *284*, 6101–6108. [CrossRef] [PubMed]

35. Cadoudal, T.; Distel, E.; Durant, S.; Fouque, F.; Blouin, J.M.; Collinet, M.; Bortoli, S.; Forest, C.; Benelli, C. Pyruvate dehydrogenase kinase 4: Regulation by thiazolidinediones and implication in glyceroneogenesis in adipose tissue. *Diabetes* **2008**, *57*, 2272–2279. [CrossRef]

36. Flachs, P.; Horakova, O.; Brauner, P.; Rossmeisl, M.; Pecina, P.; Franssen-van Hal, N.L.; Ruzickova, J.; Sponarova, J.; Drahota, Z.; Vlcek, C.; et al. Polyunsaturated fatty acids of marine origin upregulate mitochondrial biogenesis and induce beta-oxidation in white fat. *Diabetologia* **2005**, *48*, 2365–2375. [CrossRef]

37. Janovska, P.; Flachs, P.; Kazdova, L.; Kopecky, J. Anti-obesity effect of n-3 polyunsaturated fatty acids in mice fed high-fat diet is independent of cold-induced thermogenesis. *Physiol. Res.* **2013**, *62*, 153–161. [CrossRef]

38. Tordjman, J.; Chauvet, G.; Quette, J.; Beale, E.G.; Forest, C.; Antoine, B. Thiazolidinediones block fatty acid release by inducing glyceroneogenesis in fat cells3. *J. Biol. Chem.* **2003**, *278*, 18785–18790. [CrossRef]

39. Horakova, O.; Medrikova, D.; van Schothorst, E.M.; Bunschoten, A.; Flachs, P.; Kus, V.; Kuda, O.; Bardova, K.; Janovska, P.; Hensler, M.; et al. Preservation of metabolic flexibility in skeletal muscle by a combined use of n-3 PUFA and rosiglitazone in dietary obese mice. *PLoS ONE* **2012**, *7*, e43764. [CrossRef]

40. Veleba, J.; Kopecky, J., Jr.; Janovska, P.; Kuda, O.; Horakova, O.; Malinska, H.; Kazdova, L.; Oliyarnyk, O.; Skop, V.; Trnovska, J.; et al. Combined intervention with pioglitazone and -3 fatty acids in metformin-treated type 2 diabetic patients: Improvement of lipid metabolism. *Nutr. Metab.* **2015**, *12*, 52. [CrossRef]

41. Fukunaga, T.; Zou, W.; Rohatgi, N.; Colca, J.R.; Teitelbaum, S.L. An insulin-sensitizing thiazolidinedione, which minimally activates PPARgamma, does not cause bone loss. *J. Bone Miner. Res.* **2015**, *30*, 481–488. [CrossRef] [PubMed]

42. Livak, K.J.; Schmittgen, T.D. Analysis of relative gene expression data using real-time quantitative PCR and the 2(-Delta Delta C(T)) Method. *Methods* **2001**, *25*, 402–408. [CrossRef] [PubMed]

43. Kuda, O.; Brezinova, M.; Rombaldova, M.; Slavikova, B.; Posta, M.; Beier, P.; Janovska, P.; Veleba, J.; Kopecky, J., Jr.; Kudova, E.; et al. Docosahexaenoic Acid-Derived Fatty Acid Esters of Hydroxy Fatty Acids (FAHFAs) With Anti-inflammatory Properties. *Diabetes* **2016**, *65*, 2580–2590. [CrossRef]

44. Xia, J.; Wishart, D.S. Using MetaboAnalyst 3.0 for Comprehensive Metabolomics Data Analysis. *Curr. Protoc. Bioinform.* **2016**, *55*, 14.10.1–14.10.91. [CrossRef] [PubMed]

45. Cinti, S.; Mitchell, G.; Barbatelli, G.; Murano, I.; Ceresi, E.; Faloia, E.; Wang, S.; Fortier, M.; Greenberg, A.S.; Obin, M.S. Adipocyte death defines macrophage localization and function in adipose tissue of obese mice and humans. *J. Lipid Res.* **2005**, *46*, 2347–2355. [CrossRef]

46. Paton, C.M.; Ntambi, J.M. Biochemical and Physiological Function of Stearoyl-CoA Desaturase. *Am. J. Physiol. Endocrinol. Metab.* **2008**, *297*, E28–E37. [CrossRef]

47. Kuda, O.; Stankova, B.; Tvrzicka, E.; Hensler, M.; Jelenik, T.; Rossmeisl, M.; Flachs, P.; Kopecky, J. Prominent role of liver in elevated plasma palmitooleate levels in response to rosiglitazone in mice fed high-fat diet. *J. Physiol. Pharmacol.* **2009**, *60*, 135–140.

48. Yen, C.L.; Stone, S.J.; Koliwad, S.; Harris, C.; Farese, R.V., Jr. Thematic review series: Glycerolipids. DGAT enzymes and triacylglycerol biosynthesis. *J. Lipid Res.* **2008**, *49*, 2283–2301. [CrossRef]

49. Buresova, J.; Janovska, P.; Kuda, O.; Krizova, J.; der Stelt, I.R.; Keijer, J.; Hansikova, H.; Rossmeisl, M.; Kopecky, J. Postnatal induction of muscle fatty acid oxidation in mice differing in propensity to obesity: A role of pyruvate dehydrogenase. *Int. J. Obes.* **2020**, *44*, 235–244. [CrossRef]

50. Seiler, S.E.; Koves, T.R.; Gooding, J.R.; Wong, K.E.; Stevens, R.D.; Ilkayeva, O.R.; Wittmann, A.H.; DeBalsi, K.L.; Davies, M.N.; Lindeboom, L.; et al. Carnitine Acetyltransferase Mitigates Metabolic Inertia and Muscle Fatigue during Exercise. *Cell Metab.* **2015**, *22*, 65–76. [CrossRef]

51. Bender, T.; Martinou, J.C. The mitochondrial pyruvate carrier in health and disease: To carry or not to carry? *Biochim. Biophys. Acta (BBA) Mol. Cell Res.* **2016**, *1863*, 2436–2442. [CrossRef] [PubMed]

52. Herman, M.A.; She, P.; Peroni, O.D.; Lynch, C.J.; Kahn, B.B. Adipose tissue branched chain amino acid (BCAA) metabolism modulates circulating BCAA levels. *J. Biol. Chem.* **2010**, *285*, 11348–11356. [CrossRef] [PubMed]

53. Newgard, C.B.; An, J.; Bain, J.R.; Muehlbauer, M.J.; Stevens, R.D.; Lien, L.F.; Haqq, A.M.; Shah, S.H.; Arlotto, M.; Slentz, C.A.; et al. A branched-chain amino acid-related metabolic signature that differentiates obese and lean humans and contributes to insulin resistance. *Cell Metab.* **2009**, *9*, 311–326. [CrossRef] [PubMed]

54. Hsiao, G.; Chapman, J.; Ofrecio, J.M.; Wilkes, J.; Resnik, J.L.; Thapar, D.; Subramaniam, S.; Sears, D.D. Multi-tissue, selective PPARgamma modulation of insulin sensitivity and metabolic pathways in obese rats. *Am. J. Physiol. Endocrinol. Metab.* **2011**, *300*, E164–E174. [CrossRef]

55. Duarte, J.A.; Carvalho, F.; Pearson, M.; Horton, J.D.; Browning, J.D.; Jones, J.G.; Burgess, S.C. A high-fat diet suppresses de novo lipogenesis and desaturation but not elongation and triglyceride synthesis in mice. *J. Lipid Res.* **2014**, *55*, 2541–2553. [CrossRef]

56. Ranganathan, G.; Unal, R.; Pokrovskaya, I.; Yao-Borengasser, A.; Phanavanh, B.; Lecka-Czernik, B.; Rasouli, N.; Kern, P.A. The lipogenic enzymes DGAT1, FAS, and LPL in adipose tissue: Effects of obesity, insulin resistance, and TZD treatment. *J. Lipid Res.* **2006**, *47*, 2444–2450. [CrossRef]

57. Teran-Garcia, M.; Adamson, A.W.; Yu, G.; Rufo, C.; Suchankova, G.; Dreesen, T.D.; Tekle, M.; Clarke, S.D.; Gettys, T.W. Polyunsaturated fatty acid suppression of fatty acid synthase (FASN): Evidence for dietary modulation of NF-Y binding to the Fasn promoter by SREBP-1c. *J. Lipid Res.* **2006**, *47*, 2444–2450. [CrossRef]

58. Sanderson, L.M.; de Groot, P.J.; Hooiveld, G.J.; Koppen, A.; Kalkhoven, E.; Muller, M.; Kersten, S. Effect of synthetic dietary triglycerides: A novel research paradigm for nutrigenomics. *PLoS ONE* **2008**, *3*, e1681. [CrossRef]

59. Pavlisova, J.; Bardova, K.; Stankova, B.; Tvrzicka, E.; Kopecky, J.; Rossmeisl, M. Corn oil versus lard: Metabolic effects of omega-3 fatty acids in mice fed obesogenic diets with different fatty acid composition. *Biochimie* **2016**, *124*, 150–162. [CrossRef]

60. Virtue, S.; Vidal-Puig, A. Adipose tissue expandability, lipotoxicity and the Metabolic Syndrome–an allostatic perspective. *Biochim. Biophys. Acta (BBA) Mol. Cell Biol. Lipids* **2010**, *1801*, 338–349. [CrossRef]

61. Ryden, M.; Andersson, D.P.; Bernard, S.; Spalding, K.; Arner, P. Adipocyte triglyceride turnover and lipolysis in lean and overweight subjects. *J. Lipid Res.* **2013**, *54*, 2909–2913. [CrossRef] [PubMed]

62. Allister, C.A.; Liu, L.F.; Lamendola, C.A.; Craig, C.M.; Cushman, S.W.; Hellerstein, M.K.; McLaughlin, T.L. In vivo ^2H_2O administration reveals impaired triglyceride storage in adipose tissue of insulin-resistant humans. *J. Lipid Res.* **2015**, *56*, 435–439. [CrossRef] [PubMed]

63. Hallgren, P.; Sjostrom, L.; Hedlund, H.; Lundell, L.; Olbe, L. Influence of age, fat cell weight, and obesity on O_2 consumption of human adipose tissue. *Am. J. Physiol.* **1989**, *256*, E467–E474. [CrossRef] [PubMed]

64. Bernlohr, D.A. Exercise and mitochondrial function in adipose biology: All roads lead to NO. *Diabetes* **2014**, *63*, 2606–2608. [CrossRef] [PubMed]

65. Nair, A.B.; Jacob, S. A simple practice guide for dose conversion between animals and human. *J. Basic Clin. Pharm.* **2016**, *7*, 27–31. [CrossRef]

Publisher's Note: MDPI stays neutral with regard to jurisdictional claims in published maps and institutional affiliations.

MDPI
St. Alban-Anlage 66
4052 Basel
Switzerland
Tel. +41 61 683 77 34
Fax +41 61 302 89 18
www.mdpi.com

Nutrients Editorial Office
E-mail: nutrients@mdpi.com
www.mdpi.com/journal/nutrients